SPIRALS IN TIME

Also available in the Bloomsbury Sigma series:

Sex on Earth by Jules Howard
p53: The Gene that Cracked the Cancer Code by Sue Armstrong
Atoms under the Floorboards by Chris Woodford

SPIRALS IN TIME

The Secret Life and
Curious Afterlife of Seashells

Helen Scales

BLOOMSBURY
sigma

Bloomsbury Sigma
An imprint of Bloomsbury Publishing Plc

50 Bedford Square
London
WC1B 3DP
UK

1385 Broadway
New York
NY 10018
USA

www.bloomsbury.com

BLOOMSBURY and the Diana logo are trademarks of Bloomsbury Publishing Plc

First published 2015

Photo credits (t = top, b = bottom, l = left, r = right, c = centre)
P. 50 Christian Delbert/Shutterstock. Colour section, P. 1: Alex Mustard/naturepl.com (t); optionm/
shutterstock (cr); David Wrobel/Getty Images (br); IDiveDeep/Shutterstock (bl). P. 2: aquapix/
Shutterstock (t); Universal History Archive/Universal Images Group/Getty Images (br); orlandin/
Shutterstock (bl); Werner Forman/Universal Images Group/Getty Images (cl). P. 3: Andy Woolmer
(tl, tr); Borut Furlan/Getty Images (b). P. 4: All photos by Helen Scales with kind permission of Fatou
Janha and the TRY Women's Oyster Association. P. 5: Image from the Biodiversity Heritage Library.
Digitised by Smithsonian Libraries, www.biodiversitylibrary.org (tl); Mark Webster/Getty Images
(tr); University of Portsmouth (cr); Stuart Westmorland/Getty Images (b). P. 6: Georgette Douwma/
naturepl.com (t); HNC Photo/Shutterstock (cr); Alexis Rosenfeld/Science Photo Library (b);
D. Parer & E. Parer-Cook/Minden Pictures/Getty Images (cl); P. 7: Helen Scales with kind
permission of Archeotur, Sant'Antioco (tl, tr, cr); Reinhard Dirscherl/Getty Images (b);
Courtesy of Archeotur, Sant'Antioco (cl). P. 8: Laurent Charles, Our Planet Reviewed
MNHN-PNI-IRD Expedition (t); Brian J. Skerry/Getty Images (br, bl); Ulf Riebesell,
GEOMAR Helmholtz Centre for Ocean Research, Kiel (cr).

British Library Cataloguing-in-Publication Data
A catalogue record for this book is available from the British Library.

Library of Congress Cataloguing-in-Publication data has been applied for.

ISBN (hardback) 978-1-4729-1136-0
ISBN (trade paperback) 978-1-4729-1670-9
ISBN (paperback) 978-1-4729-1138-4
ISBN (ebook) 978-1-4729-1137-7

6 8 10 9 7 5

Typeset in 12pt Bembo Std by Deanta Global Publishing Services, Chennai, India

Illustrations by AARON GREGORY

Figures on pp. 55 and 56 by Marc Dardo

Printed and bound in Great Britain by CPI Group (UK) Ltd, Croydon CR0 4YY

Bloomsbury Sigma, Book Four

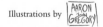

For Katie and Ruth

Contents

Prologue

Never go anywhere without your seashell. At least that was the rule that Triton lived by. He was a merman – top half human, bottom half fish – and a demigod in Greek mythology, not a fully fledged deity. Nevertheless, he did his best to dash around playing his trumpet, which was fashioned from a large shell with the end cut off; he used the trumpet's ear-splitting roar to scare off raging giants and command the seas. Triton was often outshone by his famous parents Poseidon and Amphitrite, the god and goddess of the sea, not to mention his extended family. Poseidon fathered a prodigious and eclectic assortment of offspring: there was a man-eating cyclops, a sea monster that stirred up island-swallowing whirlpools, a talking stallion, and a sea-nymph who could control violent waves and married a giant with a thousand hands and fifty heads.

Then there was Triton with his shell. His special power was perhaps not as flashy as those of some of his siblings and in-laws, but he was still someone not to mess with. One story tells of Misenus, a mortal from the city of Troy, who thought himself a gifted trumpeter and rashly challenged Triton to a musical contest. Outraged by all the boasting, the demigod shoved Misenus in the sea and drowned him. It seems Triton was a bit sensitive about his seashell trumpet.

Beyond myths and stories, seashells have always been highly valued and revered in the real, human world. Since prehistoric times, we have found shells, picked them up and looked at them in wonder. People have contemplated the seashells' beautiful shapes and the mysterious ocean realm they come from, and turned them into great treasures. For centuries, the wail of conch trumpets has echoed across the peaks of the Himalayas, calling Tibetan Buddhist monks to prayer. The conch shells inhabit the Indian Ocean and have been carried hundreds of miles inland, high into the mountains, where they are carved with intricate designs, decorated with jewels and precious metals and adorned with colourful ribbons. Standing on the rooftops of monasteries, monks play shell music into the skies to ward off approaching storms and drive away evil spirits.

Sadly, though, in more recent times, people have begun to lose this sense of awe in seashells. Their magnificence is fading and being replaced instead by inelegant clutter. I brooded on this while hunting around the internet for the words 'seashell' and 'figurine'. A cavalcade of aquatic kitsch unfolded across my screen, and one image in particular stuck in my mind: a little seashell man. His body was a large cowrie shell, his head a slightly smaller one – the opening gave him a goofy, crinkled smile – and glued on top was a cockleshell hat. His arms and legs were made from four twisted turret shells that poked out at odd angles, and he sat on an elephant made from a dead starfish with one leg raised as a trunk and clam shells for ears (not to worry, though – I'm sure it's what the starfish would have wanted). Another spectacle of shellcraft dreck – available to buy at optimistically high prices – was a series of ceramic human heads covered in dreadful jumbles of seashells along with strings of pearls, craggy antlers of dead coral and glittering rhinestone seahorses; these unfortunate mannequins looked like mermaids who'd fallen into Poseidon's treasure chest and come out much the worse for wear.

I encountered yet more seashell trinkets in a rather unexpected place. At London's Natural History Museum I was invited to go behind the scenes to the basement rooms, where their phenomenal shell collection is kept. They have millions of specimens, catalogued and neatly arranged species by species, but as I walked in the first thing I saw was a glass-fronted cupboard housing a miscellany of altogether more peculiar objects. The curators call this their 'cabinet of horrors'. It contains the various shell paraphernalia they've been given over the years; some are real shells, others are plastic replicas. Among the gubbins there are ornamental ships with sails made from scallops, and a telephone shaped like a conch shell, taking the phrase 'a word in your shell-like' to its logical conclusion after the Victorians noted that human ears have a spiralling shape similar to shells. There's a tiny shell-covered piano, and a stack of cowrie shells, each with plastic eyes and a pair of wire-rimmed spectacles that transforms them into studious turtles. Gluing a pair of wobbly eyes on a cowrie is harmless enough, I suppose, but it's a far cry from the men and women who buried their dead with shells in a sign of great respect and mourning. I'm not saying we should go back to placing shells in graves, but you've got to admit that it's funny how things change.

Even when they're not being sculpted into truly horrible ornaments, seashells have gained something of a reputation as clichéd emblems of the beach and disposable tokens of all things nautical. Lots of us live in cities – permanently tuned into the digital world and out of the natural world – so it's perhaps no surprise that when shoppers buy flip-flops studded with cowries, or shell necklaces, or lampshades made from Windowpane Oyster shells, most will have no idea where these things came from, or realise that they were made by living, wild animals.

Despite all this, there is still something about seashells that even in our busy modern lives makes many of us stop and wonder for a moment. We find them on beaches, we enjoy the feel of them in our hands, and we hold them to our ears to see if the stories are true about the sound of waves getting trapped inside. Then we take them home and arrange them on bookshelves or in the bathroom, where they remind us of a tranquil day at the coast and equip us with delicate connections to the sea. As well as being something elegant to look at, and a small treasure we found for ourselves, the shells whisper tempting questions. Where do all the shells come from? Who or what sculpts them? How are they made and, perhaps more intriguingly, why?

This book will answer those questions, and many more besides. It is my attempt to set the record straight, to throw out the novelty knick-knacks and reinstate seashells to their rightful place as glorious objects that can tell us so many things. I will show how seashells can offer us insights into the minds of our distant ancestors, and teach us about beauty and form and the curiosities of life on Earth. I will tell the stories of some of the people who have devoted themselves to shells; people who have used them in ways that make the afterlife of seashells both surprising and splendid. And I will put the animals back inside their shells and reveal the extraordinary lives of the shell-makers.

Take the seashells known as Giant Tritons, named after the Greek demigod and often used to make trumpets in the real world. Now and then they can be spotted swaggering around on coral reefs in the Indian and Pacific Oceans, their huge shells in tow; with handsome, elongated twists like polished tortoiseshell and bigger than an actual trumpet, these are one of the largest and finest of all the seashells. From under a triton's shell protrudes a single, muscly foot covered in leopard spots, a pair of yellow and black striped tentacles, and a pair of piggy eyes. Their highly sensitive tentacles probe and taste the water for the whiff of dangerous animals that plenty of other reef denizens hope never to bump into.

Crown-of-thorns Starfish are the size of car wheels and are covered in a tangle of venomous prongs and spines. They clamber up onto living colonies of coral, flop out their stomach through their mouth and digest the hapless creatures below before slurping up their liquefied remains. These starfish are formidable beasts, but they are utterly petrified by tritons. Place the starfish in an aquarium and pump in seawater that has recently washed over a triton and the normally sedate starfish will spring to life and do its best to clamber out of the tank and scram. In the wild, when a triton catches up with a Crown-of-thorns Starfish, it is somehow immune to the noxious spines. The hunter smothers its victim with its huge foot, chews a hole through its tough skin and dribbles in saliva that seems to paralyse the starfish. Then it's feeding time for the triton.

Being partial to coral-munching starfish, tritons could play an important part in keeping reef ecosystems healthy. In the past, plague-like outbreaks of Crown-of-thorns Starfish on Australia's Great Barrier Reef have been blamed on the decline of triton populations, possibly due to shell-collectors and trumpet-makers taking too many of these beautiful shells away. It's been assumed that without their predators the starfish proliferate until swarms of them are marching across reefs, leaving a trail of destruction in their wake. Certainly, there have been outbreaks of hundreds and thousands of starfish that do serious damage to areas of reef, stripping away the living, colourful tissue and exposing the bare white skeletons. In the past, some rescue attempts haven't exactly helped things when people gathered up starfish, chopped them into little pieces – to make sure they were quite dead – and threw them back in the sea. It took an embarrassingly long time for someone to point out that a whole new starfish can regrow from a small fragment, so all they were doing was giving the outbreak a helping hand. It is, however, unclear whether vanishing tritons really have been responsible for kick-starting these plagues. A single starfish meal can feed a triton for a week, so it would take a lot of tritons to keep these reef marauders in check. However, seeing how tritons make crown-of-thorns freak out, it is possible they could disrupt starfish aggregations, shooing them away and reducing their chances of successful breeding. Crown-of-thorn outbreaks could well be a natural phenomenon (the jury is still out on how much human actions are implicated, even when they aren't helping the starfish to multiply) but they are undoubtedly a problem that coral reefs could do without.

Coral reefs protect coastlines from storms, waves and rising sea levels, and provide food and livelihoods for millions of people, but they are in grave danger from numerous threats, most worryingly climate change. These vital habitats need to be as healthy as possible to give them a fighting chance of adapting and coping with the stressful modern world, and patrolling tritons are likely to play their part in a diverse, resilient ecosystem.

As we will see, the world has come to depend in many ways on seashells and the animals that make them. They perform all manner of crucial roles, from feeding people and other animals to creating habitat and providing new medicines. Wherever shell-makers dwindle or disappear, their absence leaves troublesome holes in the fabric of life, ones that are difficult or impossible to fill.

When tritons plus all the other shell-makers are dead and gone they leave behind their empty shells, which come in a dazzling variety of shapes, sizes and colours. Some are named after things they remind us of: there are sundial shells, moon shells, bubble shells, bonnet, turban, crown and helmet shells. Some shells look like vases, and some like unicorn horns. There are shells that resemble strawberries or ice-cream sundaes; others look like coffee beans; and it's easy to imagine the deep red Oxheart Clam will start throbbing and beating any minute. There is a whole group of shells called angelwings whose delicately corrugated shells might persuade the staunchest of atheists to believe that heavenly messengers have fallen to Earth. And while most shells would fit snugly in the palm of your hand, there are many that are smaller than a pinhead, and some as wide as your outstretched arms that can weigh more than a pair of newborn elephants.

There are certainly a lot of shells to choose from and this book won't tell you everything there is to know about them. This is not a shell guide or a book on how to find and identify them, although I do hope it might convince some of you to go and take a closer look. This book is made up of my choice of shell stories, ones that together paint a picture of a remarkable company of animals along with some of the more offbeat, forgotten and little-known tales of how those shells have made their way into the human world.

My own seashell story began as a little girl on beaches during family holidays to Cornwall, the tapering English county, an almost-island surrounded on all but one side by the Atlantic Ocean. With money inherited from my grandmother, we bought a damp, stone cottage in the village of North Hill, perched on the edge of Bodmin Moor. Every school holiday, including half terms, through summers and winters, we would bundle into the car and drive for four hours west. It often felt like a long way to go, and a long way from our cats and my friends. But looking back I have my parents to thank for making sure my sisters and I grew up, at least part of the time, immersed in this wild landscape.

Each day we had a choice of things to do and places to go. We could roam around the windswept, gorsey moor and scramble up to the granite peaks including Rough Tor, the highest point in Cornwall. Often we wandered down into the wooded valley that runs next to North Hill, to swing on ropes over the river, play Pooh-sticks or go searching for rabbits. And if we wanted to go to the beach we were spoilt for choice.

From our cottage it took roughly the same time to reach the craggy cliffs of the north coast and the gentler beaches of the south. My favourite was always Trebarwith Strand in the north, not far from Tintagel and its King Arthur memorabilia, which I wasn't especially interested in. I was always much more excited by Trebarwith's huge rocks that formed pools big enough to swim in at low tide, and by the dark caves, carved into the base of mountainous cliffs, where surely there was buried treasure to be found if I just kept looking for it. Not forgetting, of course, the long sandy beach that stretched into the distance. There I built sandcastles and, sticking with convention, decorated them with seashells. Best of all, I liked finding shells that were worn away on the outside to reveal the spiral hidden underneath. They seemed to me the most exotic, magical secrets, things that I had assumed were just made up – like shimmering mirages on a hot road, or double rainbows – until I saw them for myself and had to readjust my view of the world. I had always wanted to know what lived inside these neat twists and wondered if their bodies went all the way through each loop to the middle.

Occasionally I've taken shells home with me, but I don't remember ever gathering an organised or substantial collection. Instead, my shell-collecting has always been rather haphazard. Perhaps I enjoy the hunt more than the final prize. I only keep the ones I especially like the look of or that hold a special story I want to remember; I find them scattered here and there around my house, in a jewellery box or at the bottom of a pocket together with a fingernail of sand.

One year, when I was probably 13 or 14, I became fixated with painting watercolours of shells, mussel shells in particular, and I got good at rendering their fine blue and mauve lines. I remember my older sister having a large jar filled with yellow and orange periwinkles, which I always presumed she had collected herself. I loved to dip my hand into it and listen to the shells clanking around like marbles. Much later I learned that they had been Flat Periwinkles that live huddled among Bladderwrack and Knotted Wrack seaweeds, where they resemble the gas-filled bubbles that I loved to squeeze and pop.

The Cornish coasts and my childhood searches for spiralling seashells nurtured my curiosity in the wild, inscrutable seas, and almost without realising that I'd made a decision I knew I would become a marine biologist. The deal was sealed in my late teens when I began to explore Cornwall's chilly Atlantic waters from a new perspective.

After going along to a free 'try dive' session at the local swimming pool at home, my friend Helena and I both signed up to a scuba-diving course (our instructors could never remember which of us had an 'a' at the end of our name). All the way through sixth form, we spent one evening each week clambering into dive kit, jumping into the deep end and learning to be fish.

Then, in the summer holidays, we would pile our gear into Helena's ancient, sky-blue Ford Cavalier and drive down to the far west of Cornwall, sometimes breaking the journey overnight in North Hill to let the engine cool down. We camped in a field near Penzance, watched shooting stars by night and went diving every day. At first, the cold, greeny-grey waters and strong underwater currents were daunting and difficult, but it didn't take long before I felt at home beneath the waves. We snooped around old, crepuscular shipwrecks that didn't look much like ships any more, and spent hours meandering across rocky reefs encrusted with sealife. There I saw squadrons of crabs and starfish, crowds of ghostly Dead Man's Fingers (a type of coral), colour-changing cuttlefish hanging in the water like miniature submarines with rippling skirts, gardens of flowerlike anemones in reds, oranges and pinks, and a solitary Cuckoo Wrasse with brilliant blue stripes would often follow us around, as if he wanted to know what we were up to; all things new to me. And always there were seashells. I saw for myself that they aren't just beachside decorations but of course they are everywhere, scattered across the seabed – living and dead: scallops, cowries, cockles, clams, whelks. I filled my eyes and logbooks with as many of these encounters as I could, and became hopelessly addicted to the underwater world.

Following our trips to Cornwall, and clutching our dive certification cards, Helena and I both headed off to university. She studied languages and went into the wine trade, eventually moving to Australia and taking her dive kit with her. I studied ecology and marine biology and continued with what's become a lifelong love affair with scuba-diving. Besides exploring the seas wherever and whenever I could, I decided to try and do my bit to help protect the oceans and ocean life from the onslaughts of the modern world. I had seen for myself the deterioration of marine habitats and I began to notice how every creature matters, no matter how small and apparently insignificant. For many years I have lived and worked in countries around the world, investigating the problems of overfishing and working on strategies to protect the species and ecosystems in the greatest jeopardy. And all the while, throughout my research and travels, seashells have followed me around.

I have seen living shell-makers going about their lives, ambling across coral reefs or sitting still where they are and gently sifting seawater. I've marvelled at the bright colours of nudibranchs – seashells without shells – and often wondered why it is that I can't stand slugs on land, but give them a lick of colour and drop them in the sea and they become quite adorable. On one occasion, strolling along a tropical beach, I foolishly picked up what I thought was an empty seashell and got pinched by the hermit crab inside – it wouldn't let go, no matter how much I yelled at it. Now I'm a lot more wary of the animals that borrow shells.

I have also seen how people use shells and how they depend on them in many ways. In the hot, dry forests of giant baobabs in Madagascar, I found shells of African land snails (relatives of seashells) filled with rum and honey and left as offerings to the spirits of the forest. Many times I've walked through the miasma of a tropical fish market – in the Philippines, Thailand and Fiji – and seen piles of cockles, clams and other shellfish that offer a cheap source of protein for everyone. I have also witnessed the darker, elite side of shellfish. In remote fishing villages in Borneo, I saw the shrivelled meat from hundreds of illegally caught giant clams drying in the sun, destined for Asian markets where people pay top prices for these chewy delicacies.

At a fancy restaurant in Malaysia, I was offered a bowl of giant mangrove snails and had to politely decline, not because they were rare and threatened but because I couldn't quite bring myself to wrestle them out of their shells and then swallow them down. But on other occasions I've enjoyed myself much more eating shellfish, perhaps most of all on England's Norfolk coast, a few hours north from where I now live. Bags of fresh mussels are left on a table by the roadside in the village of Morston, where marshes with slurpy blue mud stretch out to the flat grey waters of the North Sea. We help ourselves and I lean in the window of the cottage, passing in a five-pound note. 'My husband gathered them this morning,' says the voice inside.

In my years of study and diving, I've also learned that there are masses of shelled animals living in the oceans that we could quite legitimately call seashells. On Cornish dives, I kept trying to bring back one of the empty, hollow sea urchin shells the size of grapefruits that I would often find lying on the sand, but every time it got broken on its journey to land. Crabs, lobsters and shrimp (including little cleaner shrimp on coral reefs and in tide pools that I've occasionally persuaded to give me a manicure) also have hardened, external shells. There are myriads of intricate sea creatures that spend their lives drifting with the currents. Most can only be seen with the aid of a microscope and they are known, generally, as plankton: foraminifera and coccolithophores sculpt chalky shells, some looking like snowflakes, others like bits of popcorn stuck together; diatoms and radiolarians make their shells from silicon and look like miniature glass Christmas tree ornaments, triangles, diamonds and stars. All of these living things have their own stories to tell and important roles in life on Earth, but this book is about just one particular group that, to my mind, are the greatest shell-makers of them all. These are the animals that go by the name of Mollusca – the molluscs.

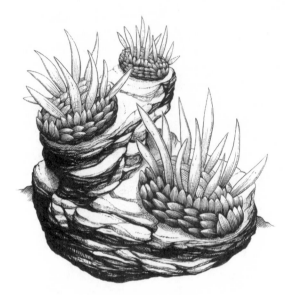

CHAPTER ONE

Meet the Shell-makers

No matter where you are in the world, you will never be far from a mollusc. These are some of the most abundant, most cosmopolitan animals on the planet, not to mention their being among the toughest, smartest and strangest creatures ever to evolve. They include familiar creatures like snails and mussels, clams and squid, as well as lesser-known varieties like chitons and nautiluses and argonauts. Molluscs (known in America as mollusks) are the animals that make seashells although, admittedly, not all of them do. Shell-free varieties exist, including octopuses and slugs as well as animals you'd be forgiven for thinking were shiny little worms. But the great majority of molluscs produce a shell of some kind. To tell tales of seashells we need to begin here, with the story of the molluscs.

No one knows exactly how many molluscs there are in total. Often-quoted numbers range from 50,000 to 100,000 known, named living species. The reason we don't know for sure is because there hasn't been a single mollusc catalogue. To name a new species, all you need do is publish a peer-reviewed paper describing it, showing why you believe it is new and hasn't previously been named, then deposit a specimen – the type – somewhere that other people can go and look at it, usually a museum. You don't have to inform some grand master of molluscs that you've found a new one, but simply add your piece of knowledge to the sprawling mountains of academic literature. And with many tens of thousands of species – including a number that have accidentally been named more than once – it's no wonder things have got rather out of hand. That's now changing, with the launch in 2014 of *MolluscaBase*, an online repository for mollusc species. It's a gargantuan effort led by teams of malacologists – the people who study molluscs – and together they are painstakingly sifting through the literature to compile a definitive mollusc list. Every year more species will be added as other malacologists venture out and delve ever deeper into the world of molluscs. Because the truth is, if you want to find new molluscs, all you really have to do is go and look for them.

Back in 1993, a group of marine biologists arrived on the Pacific island of New Caledonia with one thing on their minds. They planned to find as many molluscs as possible in one month. Led by Philippe Bouchet, from the Muséum national d'histoire naturelle in Paris, the team of scuba-divers spent a total of 400 person-hours rummaging through the deepest recesses of coral reefs in a lagoon on the island's north-west coast. They hand-picked specimens, brushed stones, cracked open solid chunks of dead coral, and even used a waterproof vacuum cleaner to carefully slurp up the tiniest hidden gems.

By the end of their expedition, Bouchet and his co-workers had gathered an astonishing 127,652 seashells.

Then the really hard work began. It took years for experts to sift through the collection and divide it into specimens that looked like distinct species (known as morphospecies) but weren't yet fully identified. That would have taken even longer.

In total they found 2,738 morphospecies. That's more than all the marine molluscs that live throughout the entire Mediterranean Sea, and almost four times more than the number found around the British coasts. And it's a higher diversity than in similar areas of habitat studied anywhere in the tropics, the most species-rich regions of the planet.

It's hard to say exactly how many of the New Caledonian molluscs were previously unknown species, because they weren't all identified. However, the team estimates that within the most diverse groups as much as 80 per cent of their shell haul was new to science. A lot of the shells they found were incredibly rare. One in five were singleton shells, that the divers found only once. If they had carried on looking, Bouchet and the team think they might eventually have taken their tally to well over 3,000 and perhaps closer to 4,000 species. That's potentially 4,000 species of molluscs, in one coral lagoon, on one tropical island.

Based on findings like these, many experts – Bouchet included – estimate that, including all the species that haven't yet been found, there could be 200,000 types of molluscs. Bear in mind that, currently, there are roughly 1.2 million known and named species on Earth and around 250,000 of them live in the sea. The only animal group more species-rich than the molluscs are the arthropods, a gaggle of invertebrates that includes crustaceans, spiders, millipedes, centipedes and the stupendously diverse insects; they alone clock in at around a million described species. Nevertheless, the molluscs take a highly respectable second place (especially as insects don't live in the sea, with a few minor exceptions that dip their toes in saltwater). It means that the insects are missing out on at least 90 per cent of the living space on

Earth (including all the vast three-dimensional space that's available in the open oceans, from the waves down to the deep seabed), which strikes me as a bit of an oversight.

There is undoubtedly a vast number of molluscs living in shallow tropical seas, but if we wanted to track down all the different varieties in the world we would need to visit a lot of other places, too. Molluscs first evolved in the ocean at least half a billion years ago, and since then they've moved into just about every available habitat beneath the waves and beyond. To find the very deepest marine molluscs we need to venture into the hadal zone, six kilometres (four miles) down, a place named after the underworld Hades of Greek mythology. This truly is a realm of fire and brimstone, and one of the most hostile parts of the planet. It's there, along cracks in the Earth's crust, that hydrothermal vents spew scorching, corrosive water from the deep beneath the sea floor, forming tall chimneys called black smokers. The only thing stopping the water from boiling is the crushing pressure. These weird, alien landscapes were first discovered in 1977 by researchers aboard the submarine *Alvin,* exploring the deep sea off the Galápagos Islands. They weren't expecting to find anything living down there at all, but in fact they saw luxurious eco-systems, including masses of shell-making molluscs.

There are sea snails living on hydrothermal vents with spiralling shells the size of tennis balls, hundreds of them squeezed into every square metre of space. For food, down in the permanent dark cut off from sunlight, they rely on colonies of bacteria that grow inside their gills, and harness energy from sulphur compounds in the water. A recent study split these snails into five species, based on differences in their DNA. To look at, they're impossible to tell apart. One of the new species is *Alviniconcha strummeri,* named as a joint tribute to the research submarine and to Joe Strummer, the lead vocalist and guitarist of the British punk band The Clash. It was a nod to these hard-as-nails snails that live in the most acidic, most sulphur-ridden hydrothermal vents in the Pacific Ocean, close to the islands of Fiji. And like many of the band's 1970s punk followers, the snails have spiky hairdos in the form of a bristly layer of protein known as the periostracum, which covers their shells.

Above the hydrothermal vents, there are molluscs that browse the vast oozy plains of the abyssal zone, and others clinging to the sides of undersea mountains. Not all molluscs in the sea are confined to the seabed, and many abandoned a creeping, sedentary life and swam off into open waters, the biggest living space on the planet. Then there are the molluscs that marched into the shallows at land's edge, burrowing down into shifting mud and sand and clamping themselves to rocks. Some molluscs didn't stop at the high-tide mark but kept on going, migrating into briny estuaries, then freshwater, and on into rivers, lakes and ponds. And a few intrepid travellers hauled their shells out onto dry land. There are molluscs that live in trees, up mountains, in sizzling deserts and in other seemingly unlikely places. A 2010 expedition into one of the world's deepest cave systems, the Lukina Jama–Trojama caves in Croatia, discovered the transparent shells of minute, blind snails that live more than a kilometre beneath the Earth's surface. Even the snails and slugs nibbling plants in our gardens are essentially ocean migrants, whose ancestors were molluscs that came from the sea. Just about the only thing molluscs haven't managed to do is take to the skies (although there are some that hitch-hike across the globe, stuck to the feet of migrating birds or lodged alive inside their gizzards).

With so many molluscs living in so many places, this raises a question. What is the key to their success? In order to figure that out, there is first a bigger, deeper question to ponder: what exactly *is* a mollusc?

'There is nothing quite like a mollusc,' wrote Colin Tudge in his book *The Variety of Life*. They are indeed a peculiar bunch, but the tricky part is figuring out exactly which features molluscs have that make them unmistakably different from everything else.

The word 'mollusc' has been around since Aristotle's time. He used it to refer to cuttlefish and octopus, and other soft things. The current use of the word seems to stem from the eighteenth-century Latin term *molluscus*, from *mollis*, meaning soft. However, going around poking animals is not much help in deciding what is and what is not a mollusc.

Over the years, 'molluscs' have included an assortment of animals, slung together because they superficially look alike. Barnacles were once labelled as molluscs and, admittedly, if you squint at them from a distance, they do look rather like little limpets (which are molluscs), but they are in fact crustaceans that have the unusual habit of sticking themselves to rocks, head down, with their legs waggling in the water. In the past, microscopic creatures called bryozoans (also known as moss animals) were bunched in with molluscs, but they have since been separated out. Brachiopods, or lamp shells, look a lot like molluscs. Their twinned shells could easily pass as cockles or clams, and yet the way they build their shells and arrange their soft innards is different enough to mean that they too have been assigned a separate group. Having a soft body and a hard hat is not enough for an animal to be considered a mollusc.

Nowadays, various animals are confidently grouped together in the phylum Mollusca, one of about 35 phyla that divide up the animal kingdom. Other phyla include the arthropods (all those fluttering, scuttling insects, crustaceans and so on), chordates (all the vertebrates plus a few strange cousins like sea squirts and pyrosomes) and echinoderms (starfish, sea urchins and sea cucumbers). The mollusc phylum is divided into eight living classes, with several more that are now extinct. By far the biggest contains the gastropods. Take a random pick of all the molluscs, and four times out of five you'll get a gastropod of some sort. These are the 'stomach feet' creatures (from the Greek words *gaster* meaning stomach and *podos* meaning foot) because they generally creep around on a single foot with a mouth on the underside. Most of them live inside a spiralling shell and are known, quite loosely, as snails. The ones that have reduced or lost their shells are known, equally loosely, as slugs (various different groups of gastropods have at one time or another lost their shells, so the things we call slugs are not actually that closely related to each other). Gastropods have evolved to live throughout the seas, in rivers, lakes and ponds, and they are the only molluscs that made it out onto land.

Clams, mussels, oysters, scallops and all the other molluscs with shells in two parts belong to the next biggest class. The bivalves have twin shells, connected by a hinge, that can open up and clamp tightly together, fully enclosing the animal inside. They live in seas and freshwaters and, along with the gastropods, they make up the bulk of species that we consider to be seashells.

The other mollusc classes are less diverse, and all of them are confined to the seas. Cephalopods are the cuttlefishes, octopuses, nautiluses and various types of squid. The name 'cephalopod' stems from the fact that in place of a single foot, these animals have a highly developed head (in Greek *kephale* means head). Many cephalopods have abandoned shell life entirely, but some have retained their hard parts and put them, as we will see, to various good uses.

Scaphopods or tusk shells are fairly self-explanatory: they look like little tusks. Often they live buried in the seabed, head down, with the tips of their shells poking out.

A little-known group of shell-making molluscs are the monoplacophorans; there's only a handful of species and they all live in the deep sea. From their shells alone, they could be mistaken for untwisted gastropods, but with multiple pairs of internal organs they are undoubtedly something far stranger. Monoplacophorans were thought to be extinct until 1952 when one came up in a dredge bucket off the coast of Costa Rica.

Chitons are a rather different class of mollusc. In place of a single or a twinned shell, they have a fringe of scales and a suit of overlapping armoured plates across their backs. You can find chitons clamped to rocks in tide pools and along the shore. They can be the size of fingernails or larger than your hand (along the coasts of the North Pacific, from California through Kamchatka to Japan, lives the biggest of them all, the Gumboot Chiton). And if they get knocked off their rock by a wave or an inquisitive human, chitons will roll up into a ball like an armadillo.

That leaves two groups of the most enigmatic molluscs, the solenogastres and caudofoveates (they are so obscure that no one has given them an easier common name). These creatures look more like worms than molluscs and none of them make shells. Instead they are covered in bristles, known as sclerites, that make them look shiny and furry.

There's no doubt that these various molluscs – the slugs, snails, squid, scaphopods and the rest – all belong together in the same phylum; their shared DNA sequences show this undeniably to be the case. Even so, the core concept of what it means to be a mollusc remains deeply contentious.

A major problem is that no one can point to a part of a modern mollusc and confidently proclaim, *See that thing right there, that's what makes this a mollusc.* If we look back into the past, down towards the base of the mollusc family tree, we should be able to see which characters have been around the longest and are therefore the most fundamentally molluscan, the things that define the group. But unfortunately, when we do that, things there aren't quite so clear-cut either.

How it all began

The oldest known fossil shells date from the Cambrian period, around 540 million years ago, with the so-called 'small shelly fossils'. This collection of minute marine fossils crops up in various places around the world. Among them are puzzling tube-like creatures that might be sponges or corals as well as masses of titchy shells, one or two millimetres (about one-sixteenth of an inch) long, that look rather like molluscs as we know them today. In the mix are shells with tightly twisted coils; some are conical like a Christmas elf's hat and some have twin shells like a clam. Most palaeontologists agree that these must have been molluscs, although a few remain cautious, pointing out that, although we have their shells, these fossils don't tell us enough about the animals that made them for us to be sure what they really were.

Alongside these tiny shelled creatures, a troupe of enigmatic unshelled animals were creeping across the Cambrian seabed. Following their discovery more than a century ago in the world's most famous fossil site, academic arguments have raged over the identity of these strange animals and whether any of them were in fact the very earliest molluscs.

On 30 August 1909, American geologist Charles Doolittle Walcott was riding his horse alone in the Yoho National Park in the Canadian Rockies when he made a ground-breaking discovery. He was looking for fossil trilobites – ancient arthropods that looked like giant, ornate woodlice – but on that day he came across some very unusual fossils. Several months later, in a letter to a geologist friend, Walcott referred to these new fossils as 'very interesting things', which was putting it mildly. In the coming years, he returned to the same spot in the Rockies many times, travelling by railway, horse and foot and in total collecting 65,000 extraordinary fossils, the likes of which no one had ever seen before. The site came to be known as the Burgess Shale.

Among his discoveries, Walcott found bizarre animals with hosepipes for snouts, terrifying creatures with massive claws and covered in enormous spines, plus all manner of shrimpy, crabby, wormy creatures that look very little like any living species. Nevertheless, he was convinced these were just strange versions of animals we know of today. In 1911, Walcott found one particular fossil at the Burgess Shale, a part of which had already been found elsewhere. Twelve years previously, Canadian palaeontologist G.F. Matthew had found a single, ribbed spine while fossil hunting in the Wiwaxy Peaks in the Rockies. He called it *Wiwaxia*. Walcott was the first to find fossilised remains of the complete animal. He decided it was a type of bristly worm known as a polychaete, a member of the annelid phylum. But it didn't have much in common with any living polychaete worms. *Wiwaxia* looked more like a slug fitted out with a suit of overlapping body armour, and with elongated knife blades sticking up in two rows along its back.

Walcott found hundreds of *Wiwaxia,* including two-millimetre-long spineless specimens and larger ones, up to five centimetres (two inches) in length. And yet, peculiar as they were, *Wiwaxia* and the other fossils found in the Burgess Shale didn't raise much more scientific interest for the next 50 years. Walcott is perhaps best remembered now as the man who didn't quite realise what astonishing things he had found.

It was only in the 1960s that palaeontologist Harry Whittington from Yale University decided to take another look. Whittington had already revolutionised the world of trilobite studies when he uncovered silica specimens, fossils made essentially of glass, that revealed dainty details of their mysterious lives. His interest in trilobites led him to the Rockies, where he reopened excavations of the Burgess Shale deposits and began a monumental task that would continue for the rest of his life.

Whittington took up a professorship at the University of Cambridge where, along with his research students Derek Briggs and Simon Conway Morris, he reassessed the Burgess Shale fossils. Together they opened a new window into the origins of animal life. It was through their work that the concept of the 'Cambrian explosion' took hold, where a plethora of complex animals appeared in a sudden flurry (although more recently the pace and duration of these changes have been questioned). Evolution seemed to be tinkering with the possibilities for life.

Among the piles of new discoveries and reinterpretations, it was Conway Morris who re-examined *Wiwaxia* and decided that it wasn't a polychaete worm after all. Inside *Wiwaxia's* mouth he found two rows of backward-pointing teeth that he thought were rather familiar. They looked to him like the rasping radula (a feature of many modern molluscs, which we will return to shortly).

While he thought the rest of *Wiwaxia's* body was too strange to win it a formal place within the mollusc phylum, Conway Morris interpreted the fossil as being a common ancestor of the group. Was this odd, spiny slug the precursor to mollusc life? Little did Conway Morris know at the time, but debates over the true identity of *Wiwaxia* had only just begun.

Since then, *Wiwaxia* has suffered from an identity crisis as people argued over whether it was a worm, or a mollusc, or something else. Nick Butterfield, also at Cambridge, waded in on the discussions early on and pushed *Wiwaxia* back worm-wards. He pointed out that *Wiwaxia's* sclerites (the ribbed scales of its body armour) were built more like a worm's bristles; what's more, its mouthparts could have been split and arranged in two parts on the sides of its head, a distinctly worm-like trait.

Wiwaxia isn't the only problematic proto-mollusc of the Burgess Shale fossils. In the original excavations Walcott found a single fossil of *Odontogriphus*, a flattened, oval creature that grew up to 12.5 centimetres (close to five inches) long, with a hardened covering across its back. It had a small, circular mouth on its underside that seemed to be adorned with radula-like chompers just like *Wiwaxia*.

Conway Morris looked at *Odontogriphus* again in the 1970s and concluded it was a common ancestor to the worms, molluscs and brachiopods. Then in 2006, after nearly 200 more specimens were found, Jean-Bernard Caron at the Royal Ontario Museum published a paper proudly claiming *Odontogriphus* for the molluscs. Caron and his colleagues also drew a close connection between these and another, even older fossil called *Kimberella*. Discovered in the 1960s in the Ediacara Hills in South Australia, the flattened egg-shaped fossils of *Kimberella* were first thought to be jellyfish. Then trace fossils were found that suggested they spent their lives not pulsing through open water but creeping backwards across the seabed, scraping up food with tiny teeth. But *Kimberella*'s teeth have never been found, so no one knows whether their snail-like scuff-marks really were made by a radula.

Some striking recent advances in our understanding of molluscan ancestry came from looking at these ancient fossils in a completely new way. For his Ph.D, Martin Smith put fossils inside a scanning electron microscope and captured images of electrons bouncing off atoms deep inside the specimens. This revealed their inner structure in unrivalled detail and convinced him that *Wiwaxia* and *Odontogriphus* were not worms. Smith worked out that both of them shed their teeth and grew new ones throughout their lives, and occasionally they would swallow them; a few fossils have teeth lodged in their guts. The bigger the animal, the more teeth it had, and each tooth swivelled relative to its neighbours. All of this, and more besides, lent weight to the idea that these fossils had molluscan kinship.

In a 2014 paper, Smith provided more support for the idea that *Wiwaxia* was an early mollusc. He studied a handful of *Wiwaxia* fossils that seemed to have a single foot, like modern slugs and snails. But part of the puzzle remains unsolved. Smith hasn't yet been able to decipher exactly where to place *Wiwaxia* on the tree of life, although he has at least narrowed things down. One possibility is that it belongs among the molluscs that don't have a single shell, the aculifera (including the chitons, solenogastres and caudofoveates). These weren't the earliest molluscs to evolve, so it would mean *Wiwaxia* wasn't a mollusc ancestor. Alternatively, *Wiwaxia* could be placed on a lower branch, as a stem group to all the molluscs. This would make it a precursor to the mollusc phylum, closer to molluscs than to any other modern group, but not *quite* a mollusc.

The concept of stem and crown groups has gained interest in palaeontological circles over the last 15 years. Crown groups are living species that share key characteristics (along with an ancestor that they all have in common, plus any extinct species that also evolved from that same ancestor). Stem groups are extinct species that have some *but not all* of those characteristics of the crown group. They are aunts and uncles to the crown group, taxonomically speaking.

This approach is helping palaeontologists to make sense of the jumble of strange animals that emerged around the time of the Burgess Shale. Many of these in-betweenie fossils could be stem groups to living phyla rather than members of fully formed phyla themselves, living or extinct. It underscores the fact that key characteristics defining a particular group of living things didn't all evolve at once but rather gradually, step-by-step, over time. It's the difference between going to a department store to buy a whole outfit compared to assembling a look from a mixture of vintage hand-me-downs, old favourites and new shoes.

Contemplating stem groups in the deep past reveals that the boundaries drawn between phyla are perhaps somewhat arbitrary. Looking at living species, it is plain to see that molluscs are very different from, say, annelids or echinoderms. But as palaeontologists peer further back through time and in greater detail, those boundaries become blurred.

If *Wiwaxia* is a stem-group mollusc, it would suggest that the radula, sclerites and a single foot were among the earlier characteristics to appear in the mollusc lineage. But it leaves an important unanswered question.

Which came first, the mollusc or the shell?

By the Late Cambrian, most of the major mollusc groups had evolved. There were indisputable bivalves, gastropods, cephalopods and chitons; scaphopods came along a while later. All of them became more abundant and diverse in the following geological period, the Ordovician. A few other mollusc groups came and went through the eons, including the now-extinct rudists; back in the Jurassic and Cretaceous these twin-shelled molluscs formed the foundations of teeming tropical reefs, similar to the coral reefs of today.

All things considered, the mighty mollusc lineage has been going for at least half a billion years, and in all that time these super-abundant, super-diverse animals have kept some secrets to themselves. We still don't really know how the different groups – the bivalves, cephalopods, chitons and so on – are related to each other, and we don't know for sure which of them came first.

Following years of research, including comparisons between living animals and more recently the arrival of genetic techniques, experts are still wrangling over molluscs. Like a pack of playing cards, the mollusc groups keep being shuffled around; should we put all the red cards together, the kings and queens in one place, should diamonds go next to hearts because they're the same colour or do they belong with the spades because they have a point at the top? Scientists keep grabbing the pack of mollusc cards from each other and moving things around.

The wobbliness of the mollusc family tree (or phylogeny) and the fact that it keeps changing shape has important implications for the way we understand evolution and the variety of life on Earth. It matters, for example, to people studying the evolution of complex brains whether cephalo-pods and gastropods are closely related or not, because both these groups have well-developed nervous systems; did these systems evolve twice, independently, or just once in a shared ancestor?

These questions, and many more, are tackled by a recent trio of studies that delve deep into the mollusc phylogeny. The three studies involved large research teams led by Kevin

Kocot from Auburn University in Alabama, Stephen Smith, now at University of Michigan, Ann Arbor, and Jakob Vinther, now at Bristol University in the UK. The methods they all used were incredibly complex, with the outcomes depending on many things, from the choice of mollusc species and outgroup (the non-mollusc species used as a comparison) to the way the data are analysed. All three teams used similar DNA sequencing techniques (using nuclear protein-coding genes, not ribosomal genes as in earlier studies), but the results they throw up don't all agree.

One conclusion that all three studies do settle on is the identity of the aculifera; they all confidently proclaim that chitons, solenogastres and caudofoveates do indeed belong together on the same branch of the mollusc family tree.

A radical outcome from one of these studies is the relationship between cephalopods and gastropods. Traditionally, these two classes were clustered together as sisters, offshoots from the same junction on the mollusc family tree. But rather than bringing them together, some of the latest genetic findings have separated the octopuses from the snails. Cephalopods could instead be more closely allied with the mysterious monoplacophorans, the deep-sea molluscs that were thought to be long extinct. Morphological studies in the past had linked these two groups, based on their fossils having a similar arrangement of internal organs, and now genetic studies have breathed new life into this idea. The gastropods are bundled, quite confidently, in with the bivalves and the scaphopods (although the scaphopods continue to be a pain in the neck to identify; we simply don't know enough about them to be sure where exactly they fit in). If this is correct then it suggests that molluscs evolved complex nervous systems on at least four separate occasions: big news for neurobiologists.

And what about the identity of the last common ancestors of all the molluscs? Did they have shells or not? This remains the subject of hot debate. Vinther and his team argue that the earliest molluscs were conchifera (the animals with single shells) and that the aculifera (without single shells) evolved later. On the other hand, both Kocot and Smith's papers keep things ambiguous: maybe it was the aculifera that evolved first, maybe it was conchifera. For now, we just don't know.

Jumping forward to the present day and casting an eye around the modern molluscs, we see no single character that all of them share, but instead there is a grab-bag of body parts; some species have them all, others only a selection. These include the radula, a muscular foot and the sclerites. Add a set of internal organs shaped like feathers called ctenidia (or gills), plus a hard shell made by a layer of soft tissue known as the mantle, and you have the basic ingredients for making all living molluscs.

This collection of mollusc body parts has proven to be incredibly malleable and adaptable. Rather than a Lego set, complete with all the specific parts to build a Star Wars *Millennium Falcon*, think of a box of modelling clay that can be made into anything your imagination allows. Similarly, each mollusc body part has been reconfigured, reshaped and repurposed over time by natural selection, allowing molluscs to wildly alter their appearance and way of life.

In effect the mollusc lineage has been riffing on a theme for half a billion years. They have been trying out experiments in how to eat and avoid being eaten, how to move about, and how to have sex and make more of themselves. This opened the way for molluscs to move into new habitats, to fill a huge range of ecological niches and ultimately to evolve into hundreds of thousands of species. Molluscs are supreme shape-shifters, and it's this versatility that could explain their roaring success, as we can see by looking in turn at each of the main body parts.

All the better to rasp/chew/stab/harpoon you with

Peer into a mollusc's mouth (preferably with the aid of a microscope) and be prepared for a terrifying show of fangs. They may be small, but they are some of the most complicated teeth on the planet.

The radula – a bristly tongue made of a protein called chitin – is covered in rows of tiny teeth, laid out across a conveyor belt that creeps ever forwards, with new teeth made at the back and old, worn-out ones falling out at the front. A single radula can have anywhere between a handful and many hundreds or even thousands of teeth, and each mollusc species has a unique arrangement of gnashers. Gastropods, in particular, have really gone to town with their teeth. They're organised into groups with names that sound like Dr Who aliens; watch out for the rhipidoglossans, the hystrichoglossans and the toxoglossans. I'd like to imagine that molluscs grin at each other to identify themselves, but of course they don't.

The precise shape and configuration of the teeth on the radula determines what molluscs can eat. Some radulas allow for quite simple but varied diets, sweeping up loose diatoms, slurping strings of seaweed like noodles or scraping at rocks and boulders covered in green slime. Limpets rasp microbes and seaweed sporelings from rocks, like a cat licking a bowl of frozen milk. The reason their teeth don't instantly shatter when they do this is because they're made from the strongest known biological material. A 2015 study found that limpet teeth, made of an iron-based mineral called goethite, are up there with the very strongest artificial materials. Limpets could chew holes in bulletproof jackets, if they wanted to. At low tide, you can see the zigzag marks they scrape across rocks and you can even hear them eating; quietly place a stethoscope on a rock near one of these little herbivores, and you should be able to make out the intermittent, sandpapery scratching as the limpet gathers its food.

Other vegetarian molluscs have evolved more specialised radulas, including the sacoglossans, a group of sea slugs that suck. They use their teeth to pierce the cell walls of plants and seaweeds, then suck out the sap inside. Many are incredibly picky eaters, feeding on a single species and, like fine gourmet diners, they have cutlery to match. Their teeth can be serrated triangles, sharp blades or shaped like wooden clogs; they're adapted to pierce particular types of underwater growth, from leathery kelp to crusty seaweeds. With their specialised teeth and diets, sea slugs divide up habitats, allowing lots of species to evolve and coexist.

Snaggle-toothed radulas become frankly terrifying in molluscs that evolved to be hunters. Many have teeth like flick knives that stand on end, locking in place during attacks, then folding safely away when not in use. A few years ago, an eerie white slug was found in a garden in Cardiff, Wales. It was a species new to science and experts had a shock when they saw its teeth: it was the UK's first predatory slug. Most land slugs, though, much to the annoyance of gardeners, are herbivores. And at a mere two centimetres (half an inch) long, the new slug is not exactly a sabre-toothed tiger, but it's no less scary if you happen to be an earthworm.

Other carnivorous molluscs have evolved more elaborate means of hunting. Cone snails, augers and turrids spit their teeth at their prey. Their highly adapted fangs are hollowed-out harpoons, which they load with a complex cocktail of deadly toxins to instantly paralyse unsuspecting worms and fish. These shells can be so toxic that they occasionally kill a full-grown human (a baffling ability that we will come back to later). There are also plenty of molluscs that have turned their radulas on their own kind. Their modified mouthparts drill neat, circular holes in shells; they then squeeze digestive enzymes into the hole and slurp out the contents. These ones are known, perhaps a little unfairly, as boring molluscs.

Not all molluscs have radulas. Bivalves lost theirs, and instead feed using their feathery gills. Most of them, including oysters and mussels, have adopted an idle approach to life. Instead of dashing after prey or crawling around looking for weeds to munch, they settle down on the seabed and stay put (more or less), and let food come to them. Tiny hairs called cilia cover their gills and beat rhythmically, creating a current. This draws oxygen-rich water inside the shell for the bivalves to breathe, and also brings in floating particles that stick to the gills in a layer of mucus. A gentle trickle of nourishment – mostly in the form of plankton – gets wafted along by the cilia, towards the bivalve's mouth. Most of them have evolved enormous gills, folded up inside their shells in a W-shape, offering a large surface area to filter food from the water around them.

Multi-tasking like this is another important factor behind the molluscs' great success. Different organs have been put to various different uses, depending on the circumstances. Gills are used to breathe and to gather food; the heart can both pump blood around the body and filter impurities from it, acting like a kidney. There are also a whole host of different uses for their singular feet.

Best foot forward
Wide sandy beaches on the Pacific coast of Costa Rica are home to sea snails that have learnt how to surf. Legions of olive snails swash-ride the waves that lap up and down the beach, using their feet as underwater surfboards; it's a more energy-efficient way of getting around compared to crawling. Once it's landed at the top of the beach, a surfing snail will put its broad, muscly foot to another use, turning it into a pouch to trap prey. Like a cat burglar with a stripy top and a sack, it engulfs its target, then quickly smuggles it away, burrowing down in the sand. And these olive snails are not choosy eaters; pretty much whatever they bump into, they will try and shove into their foot pouch. Usually it's another

olive snail, because there are a lot of them about, but sometimes they find something else. Winfried Peters, from Indiana University-Purdue University Fort Wayne, has studied olive snails and offered them a variety of potential foodstuffs. He has filmed one of these thumbnail-sized snails trying its very best to swallow a pencil.

Some molluscs, including limpets and chitons, use their feet to clamp themselves tightly to rocks and stay put (creep up and gently prod a limpet and you'll see it quickly clench down; then it becomes almost impossible to shift). Usually, though, the molluscan foot is a means of getting from A to B, often accompanied by lashings of gummy slime. So how exactly does an animal with one foot walk through glue?

The tiniest gastropod molluscs move around on hairy feet. Minute estuarine snails, called *Hydrobia*, have feet covered in masses of cilia similar to bivalve gills. These beat like a thousand tiny oars, propelling these little gastropods around their muddy homes. This method isn't powerful enough to shift larger snails and slugs, so instead they move around on waves of muscular contraction that ripple along their feet. The waves generate just enough force to slowly pull or push them along, at a speed of generally between a millimetre and a centimetre per second, in one direction only; for the most part, slugs and snails can't go backwards.

The silvery trail molluscs leave behind them as they crawl along is made of sticky stuff that doesn't play by the rules. Scientists discovered 30 years ago that molluscan slime changes its behaviour, depending on how firmly a snail or slug pushes against it. A blob of slime is indeed very sticky, but give it a squeeze – as when a wave of contraction passes by – and it turns into a free-flowing liquid. This reduces the friction on part of the foot and allows the mollusc to push forwards. Sliding through slime is an effective way for molluscs to move, to climb walls, trees and rocks and hang upside down, but it comes at a great cost; some species use up 60 per cent of their energy on making protein-rich mucus. To try and save energy, many slime-sliders including periwinkles will sniff out and follow the fresh trails laid down by other molluscs.

Clams, scallops and other bivalves don't glide around on their feet; instead they shuffle, hop, jump and dig. When it feels the need, a cockle can poke its foot out and shove itself along, hopefully out of harm's way. And while scallops clap their shells together and swim through open water for short bursts, they will also use their feet to dig down into the seabed and bury themselves. Burrowing opened up a whole swathe of new habitats for the molluscs, as did their ability to swim.

Cephalopods have highly adapted feet. Part of them has evolved into a hollow tube that squirts out water and thrusts them through the sea, by jet propulsion. And somewhere along the line, the cephalopod ancestors reshaped their feet to sprout clusters of arms and tentacles, making them the most dextrous of all the molluscs (and you can easily tell octopuses from squid by counting their arms and tentacles: octopuses have eight arms, with suckers all the way along; squid have eight arms plus two tentacles, that only have suckers at the end).

And there's little doubt that the most charming adaptation of the mollusc foot is in the gastropods that fly through the open ocean. Sea butterflies and sea angels are gastropods that bade farewell to the seabed, split their feet into two tiny wings and flitted off into the big blue yonder.

A thousand and one uses for a shell
Pulling one final piece from the mixed bag of molluscan body parts, we are left contemplating the shell. And, as it turns out, there's a lot you can do with one of these wonders of calcium carbonate.

Sculpted and moulded by natural selection, the mollusc shell has proven to be an extremely useful piece of kit. Molluscs use their hard shells, and the soft mantles that make them, to move, to eat, to hide and to fight, plus a few other surprises along the way.

Starting with the mantle, this draping cloak of tissue has various uses other than shell-making (which we will come to later). Often, mollusc mantles are quite beautiful. Cowries stick out their mantles and flap them over the tops of their shells (it's because of the mantle that cowrie shells are so shiny and smooth). In some species, the mantle offers a disguise, matching the shell brilliantly to its surroundings; some spindle cowries have bright red mantles, covered in knobbles, camouflaging them against the soft corals they live on. The shell-less nudibranchs harbour a variety of noxious compounds in their bodies, and the ostentatious colours of their mantles shout 'move along, nothing to eat here' – predators soon learn to steer clear. Cephalopods have the most sophisticated mantles of all; squid and octopuses can change colour in the blink of an eye, to communicate messages to each other or instantly blend with their surroundings, camouflaging themselves with cloaks of invisibility.

Many bivalves have rolled part of their mantle into a hollow tube, called the siphon, which they use like a snorkel. Burrowed in sand and mud, they reach up into the water to breathe and feed. Native to the Pacific Northwest, in Canada and the US, clams called Geoducks (pronounced 'gooey-ducks') have colossal siphons, up to a metre (three feet) in length. They allow the clams to live deep down in soft mud, and are so big they no longer fit inside the shells, but remain permanently stuck out, like an elephant's trunk (or perhaps an outrageously huge phallus). In China, Geoduck siphons are considered a culinary delight.

Another tube protruding from the mantle is an extendable proboscis, tipped with sensory cells. Predators and scavengers use them to sniff out things to eat (while cone snails spit teeth out of theirs). Cooper's Nutmeg Snails have an extra-long proboscis, several times their own body length, and for ages biologists wondered what it is they eat; clearly it's something the snails don't want to get too close to. The answer came after a chance encounter, when scientists from Scripps Institution of Oceanography were diving off the San Diego coast. They saw a nutmeg snail sneaking up to an electric ray, a flattened relative of sharks that can generate an electric shock to capture prey and deter predators; the jolt is equivalent to the shock from a car battery. But the nutmeg snails go unnoticed and un-zapped. They use a sharpened radular tooth at the end of their proboscis to make a small incision in a ray's belly and suck its blood. These snails are the vampires – or perhaps the mosquitoes – of the mollusc world.

Some molluscs have adapted their mantles for moving around. The marvellous *Grimpoteuthis* or dumbo octopuses (rarely are both scientific and common names so good) slowly glide around the deep sea, flapping extensions of their mantle that look like enormous ears. The mantles of cuttlefish extend into a fringe of long, narrow fins, like a frilly petticoat, that undulate in gentle waves as these cephalopods hover in the water and smoothly swim along.

As for the hard calcium carbonate shells secreted by the mantle, these are first and foremost a means of protection and a safe place to hide: a portable home. Bivalves are the best protected of all the molluscs; with their two halves closed shut, they are extremely difficult to get into, as anyone who's tried to open an oyster will know. Gastropod shells, on the other hand, tend to have a weak spot: the opening where their head sticks out. Limpets overcome this by fixing their shells tightly to rocks (they also use their shells in defence, when a predatory starfish shows up, by standing tall – so-called 'mushrooming' – then stamping down hard on the invading tube feet). Most snails can pull their heads inside their shells, and many have evolved a separate door, the operculum, which they swing shut behind them. This helps deter intruders, and prevents land-living snails from drying out.

The mantle and waterproof shell played key roles when molluscs first clambered out of the water and adapted to life on land. A cavity underneath the shell acts as a water reservoir, to see them through parched times, and part of the mantle forms a simple lung that draws oxygen from the air. Various other things go on within the safe confines of the shell, including fertilising eggs and rearing babies; instead of laying eggs and leaving them unguarded, some snails keep hold of their young until tiny, fully formed infant snails crawl out.

Plenty of molluscs use their shells not just as a place to live but as a weapon. In particular, molluscs have evolved ingenious strategies for breaking into each others' shells. There are whelks that jam the lip of their spiralling shell between the gaping shells of cockles, preventing them from closing shut, then slurp out their soft insides. Tulip shells use their tough shells as battering rams to smash their way into other molluscs. And there are gastropods that have become expert at shucking oysters; they use a prong sticking out of their shells to jemmy open hapless bivalves.

Shells have also helped molluscs adopt different modes of moving around. Chambered nautiluses use their shells as flotation devices. They're divided into gas-filled chambers, which boost buoyancy and allow the nautiluses to hover effortlessly in the water column, saving energy. Cuttlefish do a similar thing, only they grow their shells internally, not on the outside. Cuttlebones commonly wash up on beaches and are offered to pet birds (and snails) to nibble; they're not really bones but are actually the cuttlefishes' modified shells, and are light, spongy and filled with air pockets.

Burrowing molluscs often use their shells to dig, most notoriously the shipworm. Admittedly, they are wormlike in appearance, but at one end they have an unmistakable pair of shells revealing their true identity – a type of clam. Using their shells to grind wood into a network of holes, battalions of shipworms have sunk entire shipping fleets, and left piers and wharves crumbling.

There are even molluscs that use their shells as greenhouses. Heart Cockles are small, heart-shaped and pink, and can be found lying on sandy seabeds near coral reefs. Like other bivalves they sift nourishment from the water, but they also grow food inside their bodies. Colonies of photosynthetic microbes in their tissues harness sunlight to make sugars. In return for a free feed, the shells give the microbes, known as zooxanthellae, somewhere safe to live and a ready supply of light; the shells have small, transparent windows that let the sunshine in.

Perhaps the most startling example of the seashell's versatility comes from the Clusterwink Snail of Australia and New Zealand. During the daytime, these denizens of rocky shores are fairly unremarkable, small yellow shells. However, if you wait until nightfall and give one a gentle prod it will glow with a greeny-blue light. Two small spots on the snail's body shine brightly, and their shells act as highly efficient diffusers, spreading the light out and making the entire shell glow.

Why go to the effort of lighting up? It's thought the clusterwinks' beaming displays surprise intruders, which will either scuttle quickly away or fall victim to other predators that have been alerted to their presence. The glowing shells essentially act as burglar alarms.

From spades and light bulbs to life rafts, battering rams and drills, the catalogue of things molluscs do with their shells is rambling and eclectic, helping these creatures live incredibly diverse lives in many different places. Mollusc shells may come in a huge variety of shapes and sizes, but all of them are made according to the same set of basic shell-making rules for turning seawater into ceramic spirals.

CHAPTER TWO

How to Build a Shell

On the banks of the Kinta River, at the furthest navigable point inland from Peninsular Malaysia's western coast, stands the former mining town of Ipoh. Behind the bustling Chinese shophouses, white colonial town hall and railway station lies a backdrop of some seventy limestone hills, clad in forest. As visitors climb steps to the Buddhist temples perched in these green humps, or descend into the caves beneath them, they are surrounded by biological treasures, including some of the world's smallest and strangest shells.

Karst limestone formations, like the ones in Ipoh, can be seen throughout South-east Asia, from northern Vietnam through Cambodia and Thailand to the Philippines and Indonesia; they rise from the sea as idyllic islands, and poke through rainforest canopies. The limestones were formed millions of years ago by the remains of ancient sea creatures, including corals and shells. Since then, their calcium carbonate skeletons have been uplifted, then eroded by wind and rain into jagged silhouettes with giant caves inside them and underground rivers running through them.

A riot of unusual wildlife lives in these limestone landscapes. Bumblebee Bats, the world's smallest mammals, flit through the caves; blind fish crawl from subterranean ponds and out onto rocks; beetles and millipedes prosper in huge piles of bat dung; and out on the rugged hilltops roam troops of leaf monkeys, including such incredibly rare species as the Delacour's Langur with its striking black and white fur (the Vietnamese name for it, *vooc mong trang*, means 'the langur with white trousers'). The chalky soils are also a haven for molluscs that find a plentiful supply of the principal raw material to make their shells.

A single Malaysian limestone hill can be home to between 40 and 60 species of tiny microsnails, each one a millimetre tall, and all of them with highly ornate shells. Of those, two or three species could be unique to that individual hill. As well as snails, there are heaps of other endemic species here, ones that are found nowhere else on the planet: geckos, crickets, orchids, begonias and spiders. Just like oceanic islands, the limestone outcrops are isolated dots of habitat where evolution dances to a different beat, generating new and peculiar species.

When biologist Reuben Clements went snail-hunting in the hills of Ipoh, he discovered a shell like no other. To find it, all he had to do was take a few scoops of soil, place them in a bucket of water and wait for the empty shells to rise to the surface (for a long time only recently dead specimens were found, and no living snails). Seen under a microscope, these tiny shells reveal their curious physique. They look like the corrugated pipe of a vacuum cleaner that's been left tangled on the floor, with the end flared out like a tiny trumpet. These shells twist and turn, this way and that, as if they can't decide which way to grow.

A few years later, Clements' colleague Thor-Seng Liew
finally tracked down live specimens of this tiny snail and set
about studying them for his Ph.D, devising a theory to
explain their bizarre coiling shapes. Liew suggested that the
snails are doing their best to avoid getting eaten by predatory
slugs. Retreating into their shells, the snails force their
attackers to reach into an empty, bendy tube while their
prospective dinner cowers at the end. The slugs' proboscis
simply can't reach into such a deep and convoluted recess.

Meanwhile, Clements and other limestone enthusiasts
have been campaigning to protect the remarkable but
often overlooked places these snails come from. Being
useless for agriculture or development, limestone hills
were left more or less alone for a long time, but now
cement companies are getting in on the act, razing them
to the ground for the limestone inside. These are imperilled
arks of biodiversity that few people have heard of. Year on
year, hundreds of species are going extinct, most of them
before they are discovered, when the hills they once lived
on are taken away.

Compared to Clements and Liew's bizarre find in the
Malaysian hills, most shells are far less erratic in the way they
grow, and indeed they are often quite predictable. For
centuries, many great minds have contemplated the elegant
sculptures and patterning of shells and wondered what might
govern their construction. They have hunted for clues to
explain the amazing realities and tempting possibilities of
shells; they have probed ideas of what makes a shell work
and which shapes may ultimately never show up; and they
imagined that if they could find ways of drawing shells, if
they could mimic what nature has been doing for eons, it
would not only bring them closer to understanding how
molluscs make their intricate homes, but they might also
catch a glimpse of the origins of beauty itself. What many
generations of mathematicians, artists, biologists and
palaeontologists have found is unexpected and elegant: to

construct an elaborate seashell – and decorate it – requires
only a handful of rules.

Of all shell shapes, one of the simplest and most pleasing is
the spiral of the chambered nautilus. The internal twist of
these ocean-wanderers is revealed when their empty shells
are sliced in two, from top to bottom. Trace the outer edge of
a nautilus shell and you'll see that it spins inwards in a very
particular way. This graceful curve was among the first shapes
in nature to be granted its own mathematical formula.

In the seventeenth century, French philosopher René
Descartes composed a simple piece of mathematics for drawing
a shape called the logarithmic spiral. Unlike an Archimedean
spiral, which has whorls that are always spaced the same width
apart, like a coiled snake, the gaps between successive whorls
on a logarithmic spiral get increasingly wide. Logarithmic
spirals flare open as they get bigger, just like a nautilus shell.

A chambered nautilus shell cut in two, revealing its logarithmic spiral.

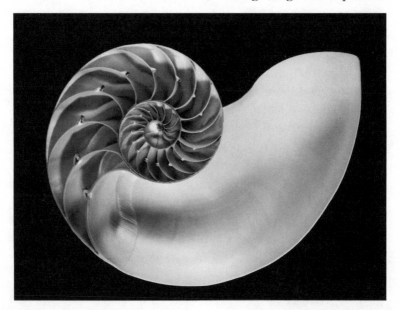

It was the cleric and mathematician Reverend Henry Moseley who, in 1838, first pointed out that many coiled shells are versions of the logarithmic spiral. Take a photograph of a nautilus shell cut in two, overlay the outline of a logarithmic spiral and, given the right dimensions, it should be a good fit.

These expanding spirals pop up all over the natural world; you can spot them in patterns of seeds in a sunflower, in spiralling galaxies, in the bands of rain and thunderstorms that swirl around the eye of a tropical cyclone, and in the path taken by a doomed moth as it flies mesmerised towards a candle. All these spirals are subtly different; what unites them is the fact that they all get bigger at a constant rate. In other words, the gaps between successive coils get wider by the same amount each time the spiral makes a complete turn around its central point. This means that no matter how big the spiral becomes, its overall shape doesn't change – and that is one of the key rules for making a seashell.

The way molluscs make shells is reminiscent of the ancient practice of coiling pottery. For thousands of years, people around the world have rolled strips of clay between their hands and coiled them into simple pots. In a similar way, a mollusc creates its shell as a hollow tube. The mantle (the fleshy cloak that spreads across a mollusc's body) lays down new shell only at this open end, known as the aperture. It does so by first secreting a scaffold of protein, which is then shored up with calcium carbonate in one of two varieties (and sometimes both): aragonite or calcite, the latter being a more stable form. The main building blocks for the shell are carbonate ions, consumed in the mollusc's diet or absorbed from seawater, and squeezed into a small gap between the mantle and the growing edge of the shell. Finally a layer of nacre, or mother-of-pearl, is added on the inside, creating a smooth layer that protects the mollusc's soft body. As this composite tube grows it becomes wider at the open end, transforming it into a cone. The mollusc scrolls this cone round and round, forming, in cross-section, a logarithmic spiral.

Before we go any further, I should point out that molluscs are not mathematicians. They aren't aware of the arithmetical elegance of their homes. These patterns simply emerge from the way they grow. You could do the same thing with a tube of toothpaste, albeit a modified version in which the opening can be made wider as you go, so that it produces an expanding cone of minty freshness. Squeeze a coil of toothpaste onto a flat surface and you should see a seashell-like twist appear (our toothpaste coils are solid, unlike the molluscs' hollow shells). Various different types of toothpaste shell will be made depending on how hard you squeeze and how quickly you move your hand away from the centre of coiling. You can make tightly coiled shapes or expansive ones that veer off and swiftly become enormous. However, keeping the toothpaste on a flat surface limits the types of shell you can make. As you've no doubt noticed, mollusc shells are not generally flat: most of them explore a third dimension.

'The problem is one not of plane but of solid geometry,' wrote Sir D'Arcy Wentworth Thompson in his classic *On Growth and Form*, as he began to grapple with the idea of three-dimensional seashells. While the First World War raged, the professor from St Andrew's University in Scotland wrote more than a thousand pages packed with his ideas of how mathematics could explain shapes in nature, from horns and honeycombs, beaks and claws, to dolphins' teeth and the shape of a splash. Thompson brought together many of his predecessors' theories about the geometry of shells, including those of Sir Christopher Wren, who mused on their architectural beauty. Some earlier shell thinkers rejected the idea of logarithmic spirals, saying they were too simplistic, but Thompson underlined their importance and brought out lots of new examples of shells that were a good fit to this expanding curve. He then set out to find a way of drawing accurate three-dimensional model shells. His central idea was that coiled shells follow a set of rigid mathematical laws, which all stem from the fact that infant shells are simply smaller versions of their future, grown-up selves.

Molluscs only ever make a single shell, but it's one they'll never grow out of. Other creatures with hard exoskeletons tend to do things differently. Crabs, lobsters and all their crustacean relatives break out of their shells every now and then, cast them aside and grow a new version, one size bigger and sometimes in a wildly different shape from the one that came before. Turtles make their shells on the inside by modifying bones in their ribs and pelvis. By contrast, molluscs make their shells on the outside, and they hold on to them. They are among the few animals on the planet that wander around carrying with them the same body armour they had as babies; the pointy tip or innermost whorl is the mollusc's juvenile shell. Day by day, the mollusc shell slowly expands, making room for the soft animal growing inside.

Thompson visualised the growth of a spiral shell as a two-dimensional shape spinning through three-dimensional space around a central axis; if you imagine poking a needle through a coiling seashell from the tip towards the open end, so that it rotates like a spinning top, then the needle takes the position of the axis. Our toothpaste shells stayed in the same flat plane and didn't make much use of that axis. Now, imagine what happens if the toothpaste sets solid straight from the tube. You can drop the spiral downwards, along that vertical axis, and create a three-dimensional coiling shell.

Taking this idea (although using paper and pen rather than quick-setting toothpaste), Thompson devised a shell-making model based on four rules: first, the cross-section of the coil must stay the same shape, but grow bigger over time (in other words, slice across our expanding toothpaste cone at any point and you will see the same shape, in this case probably a circle); second, the shell's curve expands from the centre at a fixed rate (making it logarithmic); third, the amount of overlap between successive whorls stays the same; finally, and most difficult to visualise (so don't worry too much), the angle between the spinning whorls and the central axis also remains the same.

These four rules were all Thompson thought were needed to draw any version of a coiling seashell. But what would all those shells look like? That was the question asked by another scientist who, 40 years later, was inspired to customise Thompson's model and use it to create a shell collection like no other.

The imaginary museum of all possible shells
Standing in the corner of a huge room, you see white walls stretching out in front of you and towering upwards, disappearing as if into the clouds. Suspended in the room's cavernous space are what at first glance seem to be thousands of glass light bulbs. They dangle in neat rows and columns, beginning just above the floor and reaching up way above your head. Take a closer look and you'll notice that they aren't in fact light bulbs but intricate models of seashells.

Despite their glassy appearance, these shells are quite tough and you can push through them without breaking them. As you do you see that the shells differ subtly from one to the next. As you look upwards, the shells gradually become squatter and fatter. Walk forwards and the shells at eye-height flatten out until they are no longer coiled but more flattened, like clams. Stroll on through the museum of all possible shells and you'll spot both familiar shells and some less familiar shapes.

The architect behind this imaginary museum is palaeontologist David Raup, from Johns Hopkins University in Maryland. In the 1960s he took Thompson's shell-making model and made a series of adjustments, replacing the original rules with four of his own.

First, Raup defined the rate at which a shell flares outwards. This is the whorl expansion rate, or 'W': tightly coiled shells have lower W values compared to more flared,

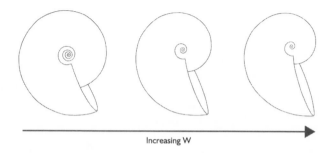

Increasing W

W: Whorl expansion rate (becoming more clam-like).

open shells. Clams and other bivalves have such high W values that they flare right open before having a chance to do much coiling. They may not look it but, in essence, bivalve shells are still spirals.

Next comes 'T', a factor that determines how tall the shell will be (the T in fact stands for Translation, meaning how much the growing spiral travels along its central axis, but it could just as easily mean Tall). In shells with a tall spire, the coil creeps downwards along the axis as it spins round and round. The further it creeps, the taller the shell spire and the greater the value of T.

Raup kept one of Thompson's rules. He admitted that in the real world, the cross-section of a shell's growing cone can change, but to keep things simple Raup fixed his shape and made it a circle. He allowed the circle to get bigger as a shell grows but it always stayed the same shape.

T: Translation rate (getting taller).

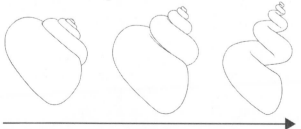

Increasing T

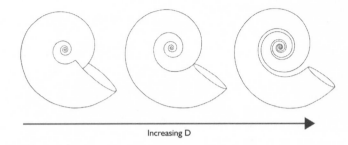

Increasing D

D: Distance from axis (becoming more wormy).

The final part of Raup's model is the distance 'D' from the whorls to the axis. Adjusting the value of D can produce thin, wormy shells with big gaps between whorls, or chubbier shapes in which the whorls touch or even squash into each other.

Armed with his new idea for plotting shell shapes, Raup did something unusual for a palaeontologist at the time: he used a computer. He bought himself time at the helm of the fastest one available. The enormous mainframe IBM 7090 was a state-of-the-art scientific computer that was intended for the design of missiles, nuclear reactors and supersonic aircraft, but for a short time Raup channelled its power into making shells. He plugged in a few combinations of values for T, W and D and programmed the computer to draw the corresponding shells on a Calcomp x-y plotter. The computer's output was a series of dots outlining several shells in cross-section, which were then interpreted into three dimensions by an artist. These drawings appeared in Raup's 1962 paper in the journal *Science*. To make more of these scatter plots would have taken way too long and been too expensive on computer time. Raup knew that the available technology was limiting his work.

His next step was to team up with an electrical engineer, Arnold Michelson, and together they tried a more affordable set-up, the PACE TR-10 analogue computer (which despite its gargantuan size was one of the earliest desktop computers, although presumably only for a rather large desk). They plugged in a wide range of values for T, D and W from Raup's shell model, hooked the computer up to an oscilloscope that traced the shapes of shells as fast-moving circles across the screen, then stood back and watched.

Unfolding before their eyes was a stunning collection of shells that looked like thousands of tiny X-rays. Among the PACE TR-10 output were examples of just about every type of shell, from the nautilus through to all sorts of coiled snails and even flattened clams and scallops. Their findings were so stunning that one of their virtual shells made it to the front cover of *Science*. Raup and Michelson had shown that from a simple set of rules emerges the great complexity of real shells. And their imaginary shell collection contained plenty more besides. Once they had traced out the shapes of all these shells, Raup turned his attention to the next big question: which of these shells can be seen in the real world?

Back inside the virtual museum of glass seashells, we can now understand how things are arranged: along one wall the thin, wormy shells become chubbier, as values for D shift from zero to one; in another direction the shells become gradually taller, as the values for T start at zero and run along to four; and from ceiling to floor, shells have progressively higher W values, from one to a million, and snails morph into clams. The glass shells are versions of the output from Raup and Michelson's PACE TR-10 computer program of all possible shells.

Now something changes in our museum. The main lights are dimmed and individual glass shells, here and there, begin to glow (funnily enough, like light bulbs). These are the models that closely resemble real shells, living or extinct. And as parts of the room light up, something becomes obvious: large regions of the museum remain in darkness.

I illuminated the real species among our glass shell models and Raup did a similar thing on paper. He plotted a graph with three axes (for D, T and W) and shaded in areas where real mollusc shells can be found, as well as the brachiopods, which are only distantly related to molluscs but even so make similar shells. Drawing in the boundaries of reality onto his imaginary shell museum, Raup immediately saw that only a small fraction of his theoretical shells have ever actually evolved. Substantial regions seem to be out of bounds. He theorised that some of the empty space in his museum was filled with 'bad' shells that, in reality, don't work. Maybe they would be too heavy or too weak, or would leave their inhabitants in some way vulnerable to attack? There is an empty region filled with shells that suffer from what Raup referred to as the 'problem of bivalveness'. Clams, mussels and scallops would be permanently clamped shut if their gentle whorls overlapped (the only option for opening up would be to build a new hinge on the outside and keep moving it as the mollusc grows bigger, and bivalves in the real world don't do this).

Other parts of the museum were empty, Raup suggested, simply because the process of natural selection hasn't got around to filling them yet. He thought that as soon as a situation arises in which those theoretical shells become useful and confer an advantage on their owners, then, sure enough, they will evolve. Other researchers disagree. They think the vacant spaces of the museum will never be filled, because the necessary genetic mutations to make those shells haven't happened and maybe never will. Their view is that natural selection doesn't have at its disposal all the genetic variation that is necessary to fill every part of the imaginary museum. Debates still rage over who is right.

Following on from Raup's original concept, many other museums of possible creatures have been built (they are now known technically as theoretical morphospaces). There are museums for beetles, aquarium tanks for fish, sea urchins and phytoplankton, even a herbarium for plants and aviaries for birds and pterosaurs. Just like the shell museum, these rambling spaces are filled with both real and imaginary beasts, and they are encouraging biologists to think about which forms and shapes in nature are possible and popular, and which are impossible or for some other reason have never occurred or will never occur.

Throughout his papers, Raup was always careful to point out that his model isn't perfect, and it doesn't account for all the things we see in the real world. For one thing, he confessed to being overly simplistic about fixing the various shell dimensions throughout a mollusc's lifespan; there are real shells that seem to shift the values of T, D and W over time, so they hop around the imaginary museum as they get older. And, as Clements and Liew found with their strange microsnails, there are some molluscs that break all the rules. One of the tiny snail species from the limestone hills of Malaysia makes a shell that spins around not just a single axis but four: the most of any known shell.

For the sake of simplicity, other features seen on many real shells are also omitted from the museum of all possible shells. For example, Raup left out the ornaments – spikes, knobbles, ribs and spines – that molluscs use to decorate their shells.

Why shape matters

Geerat 'Gary' Vermeij has probably spent more time than anyone else thinking about the shapes of seashells. Born and raised in the Netherlands, the first shells he encountered were what he describes as 'drab chalky clams' on windswept North Sea beaches. Then, in 1955, his family moved to Dover, New Jersey, where Vermeij experienced something of an epiphany. His fourth-grade teacher, Mrs Colberg, decorated the classroom windowsills with dozens of shells she had gathered during holidays to southern Florida's tropical shores. They were nothing like the shells Vermeij had got to know in Europe, being elegantly sculpted and covered in prickles and bumps. Her cowrie and olive shells ∽

were so shiny he was sure someone had varnished them.
When a classmate brought shells from the Philippines to
'show and tell', Vermeij saw these were even more exotic
and enthralling. He resolved to begin collecting his own
shells and to find out as much as he could about them.

A decade or so later, Vermeij graduated with a Ph.D from
Yale University, and since the 1980s has been Professor of
Paleoecology at the University of California, Davis. It
became his lifelong passion to understand how and why
shells grow in so many different forms throughout space
and time. He has travelled the world exploring the coasts
and seashells of nearly every continent, and published more
than a hundred scientific papers and four books about shells
and evolution. And, since the age of three, Gary Vermeij has
been blind.

Using his finely tuned sense of touch, Vermeij studies
shells by turning them over and over in his hands, feeling
their intricate shape and noticing details that other people
miss. In his book *A Natural History of Shells*, he writes about
how his hands have allowed him to explore the way shells
from different places vary in appearance: the geography of
shape.

He describes how the shells he finds on tropical shores
are radically different from those on Dutch beaches. For
starters, they are much more carefully made. Individuals
from the same species of tropical mollusc will make shells
that are identical copies of each other. They stick closely to
a set of hidden rules, imposed perhaps by the presence of so
many predators and competitors. Slightly wonky shells just
won't cut it in the race for survival in these crowded, species-
rich waters; they might not be strong enough, or well
protected enough from attack. In cooler and deeper waters,
where life in many ways is more relaxed and less extreme,
molluscs can get away with being less finicky about their
shells. On the whole, away from the tropics, molluscs are
built relatively sloppily.

Vermeij also writes in his book about another key moment in his life, when a big idea hit him. He spent the summer of 1970 in the western Pacific Ocean, on the island of Guam, on a field trip with his friend Lucius G. Eldredge. On one particular day they were searching for shells in the falling tide at Togcha Bay on the windy side of the island when Eldredge (known as Lu) handed Vermeij the shell of a Money Cowrie with its top sliced clean off. Lu made an offhand remark that he often saw crabs cutting open cowries in his aquarium tanks.

Until then, Vermeij hadn't paid much attention to the fact that he often found masses of broken shell pieces on tropical beaches and he suddenly got to thinking about predation. He realised that tropical seashells have a really hard time with so many predators trying their best to crack, smash, peel open and drill into them. He began to wonder how their shells have evolved to ward off these attacks, and soon realised there are many reasons why shape matters.

An obvious way a mollusc can avoid getting eaten is by making a very big, thick shell, but that comes at the cost of having to make and then drag around a massive, heavy lump. A more economical way to make a shell more difficult to handle and swallow is to give it a covering of spines and bumps. Realising this, Vermeij finally understood why Mrs Colberg's Floridian shells, and so many other tropical species, have fancy ornaments. In the crowded tropics, molluscs are doing their best to survive. As they grow, they can add embellishments to their shells; prongs can be added at regular intervals, or they can form a dense tangle like the quills of a porcupine. *Spondylus*, for example, the thorny oysters, are industrious spine-makers, expertly producing new ones and fixing any that have broken at a rate of a few millimetres every day.

Vermeij also figured that the pleats and corrugations on many tropical shells are a cost-effective way of creating a strong body armour that's difficult to break into while keeping the weight down. Thickening and flaring out the aperture of shells is another way of deterring predators, as in the Malaysian microsnails with their trumpet-shaped mouths.

Shape can also help shells to hide. Sleekly shaped molluscs can slip silently through the water without sending out telltale ripples that predators detect; being more hydrodynamic also allows for a quicker getaway. We can surmise that parts of Raup's imaginary museum may remain empty of real shells simply because they are not streamlined enough.

For shells that live in sandy, muddy places, shape can mean the difference between resting on top and sinking in. Epifaunal species are ones that have adapted to a life of lying on the surface of the seabed; their shells are often wide and flat, acting like snow shoes. They include species like the Big Ear Radix, a gastropod that lives in lakes across Europe; throughout their lives they continually expand a winglike flap on their shells that prevents them from sinking into silty mud. Another strategy used by epifaunal species is what Vermeij describes as the 'iceberg habit'. Instead of lying on the surface they allow themselves to sink in slightly so that most, but not all, of the shell is submerged. Scallops commonly have a curved lower shell that sticks a short way into the mud.

Shape also matters for infaunal species, those that spend their lives burrowed down into mud and sand. Among the sea snails and bivalves there are champion diggers that use their feet as spades to bury themselves completely in under a second. Some have tiny ratchets on their shells to prevent them slipping backwards, and others have smooth whorls to make sure sand and mud don't stick to them and increase the load.

Burrowing shells face the additional problem of being unearthed. If you've ever stood barefoot in lapping waves on a sandy beach, you may have noticed the sand being scoured from around your toes. When waves and currents flow around a solid object they stir sand grains into suspension and whisk them off elsewhere. To overcome this, burrowing shells evolved spines and ribs that trap sand particles and stabilise the sediments around them. A group of typical diggers are tower shells, which look like little unicorn horns; their sculpted whorls help to hold them in place in their sandy, muddy homes and reduce the chances of being swept away.

Back inside Raup's imaginary museum of all shells, there is another perplexing detail that needs explaining: all the coiling shells twirl in the same direction. Suspended from their wires, the glass models have their tips pointing downwards and their apertures all open to the right. Or, seen from the top, they coil in a clockwise direction. Raup could easily have filled his museum with shells that twist the other way, or perhaps made two giant rooms that were mirror images of each other. But he didn't, and for good reason.

Take a look at any real, spiralling shell and see which way it turns. Go and find that seashell sitting on a bookcase, or pick up a snail from your garden or local park; your shell almost certainly coils to the right. There is a smattering of species that always coil to the left, and occasionally sinistral oddities will occur in a right-coiling species, but currently the natural world favours righties over lefties. More than nine out of ten coiled shells today are dextral (curiously, a similar proportion of people are right-handed).

Shell collectors go crazy for rare sinistral specimens, so much so that over the years clandestine trades have prospered in fake lefties. Some are right-coiling shells that have undergone a bizarre molluscan version of plastic surgery, with some bits cut off and others glued back on; X-rays show their insides are in fact dextral. There are also true left-coiling shells that masquerade as something more special. Around the world, Hindus and Buddhists are summoned to prayer by the call of sacred conch-shell trumpets, known as *shankh* in Sanskrit. These are made from a large species of

Indian Ocean gastropod, known in English as a chank shell, which normally coils to the right. Rare left-coiling specimens are highly revered, and are referred to variously as *dakshinavarti shankh* or *sri lakshmi shankh*. Their anticlockwise whorls are said to mirror the passage of the stars and sun across the heavens, and the curly hair and twisting bellybutton of the Buddha. Unscrupulous shell-traders make counterfeit *sri lakshmi shankh* shells from a different species, the Lightning Whelk, which lives in the Gulf of Mexico and normally coils to the left.

A famous left-handed shell was drawn by Rembrandt. He portrayed a Marbled Cone Snail which, like most of the poisonous cone snails, naturally coils to the right. Art historians speculate that Rembrandt hadn't made a mistake, as many early shell illustrators did. Failing to appreciate the significance of coiling direction, artists would commonly etch what they saw into metal plates; their shells would then become reversed as mirror images in the printing process. In Rembrandt's case, though, it's thought he reversed his shell on purpose, for aesthetic reasons: he just felt it looked better that way. Pleasingly, other artists who copied Rembrandt's cone did so directly and faithfully, without thinking to reverse the etching, so these printed shells were restored to their rightful place as right-coilers.

The abundance of right-coiling shells in the natural world, and lack of left-coilers, comes down to one simple but inescapable truth: if right- and left-coiling snails try to mate, their genitals don't match. Not only are shells coiled one way or another but the rest of the snail's body is also asymmetrical. Female snails have a genital pore offset to one side into which a male will inject sperm through his penis. Most gastropods in the oceans have separate sexes – they are shes and hes; land snails are commonly hermaphrodites, each one with both bits of equipment, but they will pair up and take turns being male and female. Face-to-face is a popular position for snail sex, and for this to work it's crucial for the female pore and male penis to overlap: this only happens if both snails coil in the same direction (a little like when you go to shake someone's hand – it only works if you both offer the same hand). The shells and bodies of left- and right-coiling species are mirror images of each other. Even the corkscrew-shaped penis of the Asian Trampsnail twists the other way in lefties, and the choreography of their circular mating dances is reversed. In a tryst between right- and left-coiling snails, everyone is confused, and everything is in the wrong place.

To gauge just how much of a problem coiling direction is in mating molluscs, researchers place pairs of mismatched snails together in cosy containers. Roman Snails, known and eaten in France as escargots (and highly protected in England), are often used in these sorts of sex studies because most of them are right-coiling, but once in a while a lefty shows up. No matter how much the left-right partners are feeling in the mood, the slurp of a baby snail's feet never issues from the mating cubicles.

An alternative mating tactic adopted by some snails is for one to clamber up from behind on the shell of the other. Similar snail-in-a-box studies show that shell climbers have more success in crossing the left-right divide than face-to-facers, but things are still rather awkward. Far fewer offspring will result from a right-left union than from snails paired up with same-shelled partners.

All of this means that for sinistral snails in a mostly dextral world, life can be lonely. It's not that right-coiling shells are inherently any better than their left-coiling brethren, it's really just a matter of chance. Whichever form is less abundant within a species will be less likely to find a matching mate and therefore not as successful at passing on its genes; this pushes a population towards one dominant coiling direction. It just happens that at the moment right-handed shells are most abundant and get the best chances to mate. But that hasn't always been the case, and the fossil record shows that fashions can change, although exactly why this happens remains a mystery. In *The Natural History of Shells*, Vermeij describes the eight or nine ancient groups of cephalopods that, through time, evolved right- and left-coiling shells, with no particular inclination towards twisting one way or the other.

It is tempting to link the coiling of gastropod shells to the fact that when they are very young, their soft bodies also undergo a major twist. This process, torsion, is unique to the gastropods and involves all the major organs spinning around 180 degrees (clockwise in sinistral and anticlockwise in dextral shells). Among many things that move, the anus shifts to a new position above the mollusc's head. Torsion is genetically determined, but a separate gene deals with shell coiling. It is an ancient gene, known as a nodal, that evolved long ago and today governs the asymmetry of many animals, including humans: we wear our hearts on the left thanks to the same gene that makes snails twist one way or the other.

Looking back into the fossil record, there are lineages of gastropods that over time have untwisted their shells, like limpets, until they look like conical Asian hats. In at least one group, molluscs have unwound their shells, then around 100 million years later, against all the odds, their descendants have coiled themselves back up again. These changes would have been driven by mutations in the coiling gene.

Given that a single mutation in an inherited nodal gene can switch a snail from being dextral to sinistral, all in one go, it raises the interesting possibility that a new species could instantly evolve. The mating struggles that take place between mismatched shells create exactly the kind of barrier that can subdivide populations and allow new species to split off, in this case leading to separate right- and left-coilers that can't interbreed. And there are a few spots on the planet where having a rare sinistral shell can put a snail at a distinct advantage.

Satsuma snails live in the Ryukyu archipelago in southern Japan and a surprising number of them are left-coilers. It just so happens that these islands are also the realm of Iwasaki's Snail-eating Snakes. A land-snail expert from Kyoto University, Masaki Hoso, studies these snails and has spent many hours watching what happens when a snake sneaks up on a target, sliding up silently and swiftly striking from behind. Because of the way their mouths are shaped, the snakes can grasp a shell with the upper jaw while plunging their teeth through the aperture and into the soft flesh inside – but only in right-coiling snails. When they try the same thing on left-coiling snails, the snake can't get enough purchase and the shell pings off to safety. Snakes pose such a terrible threat for satsuma snails that when young dextral snails are attacked, they voluntarily amputate their feet (geckos do a similar thing, dropping their tails to confuse predators while they dash off and make their escape). Hoso has never spotted a sinistral satsuma resorting to such a risky escape strategy; they always hold on to their feet.

Mapping out the distribution of snails and snakes, Hoso found that left-coiling species of satsuma snails only occur in or near areas where there are also these fearsome reptilian predators. So it seems that avoiding the chomp of lopsided snake jaws gives the left-coiling snails the edge over right-coilers and as a consequence sinistral snails have flourished. Although it will probably be only a matter of time before the snakes likewise evolve to become left-handed.

When nature is allowed to play

The final flourish in the process of shell-making is where molluscs are at their most creative. As well as forming intricate shapes, shells are also decorated in elaborate patterns. There are few other animals that paint themselves in such a profusion of complex markings. With their spots, stripes, waves, zigzags and triangles you could perhaps assume molluscs are simply playing with their shells.

There are two strange things about the shell patterns. First, no one knows which pigments molluscs use to paint their shells. So far, only a broad group of organic molecules has been detected, including porphyrins and polyenes. The closest anyone has come to pinpointing an actual shell pigment is a carotenoid in the yellow rings of Money Cowries.

The second peculiar thing about seashell patterns is that often they go completely unseen. Many ornately painted bivalves and gastropods spend their lives hidden out of sight, burrowed in sand or mud. And there are some that grow a layer of protein (the periostracum) over the outside of their shell, often making them look like weedy rocks. What purpose, then, can there be for these shrouded shell patterns? Why should these highly decorated shells get all dressed up with nowhere to go?

For a long time, biologists assumed that shell patterns don't really matter, one way or another. The assumption was that since their output is never seen, the processes that lay down intricate patterns in a snail's shell had become unshackled from the strict forces of natural selection, and were essentially neutral – they had been left to wander around an art gallery of all possible patterns, without any rules telling them what they were allowed to do.

Exactly how and why such elaborate patterns evolve, with apparently no purpose, does seem at first to be a bizarre and inconvenient mystery, the sort of thing that creationists leap on as proof that it was God who made it so. But as scientists have unpicked the process that leads to these patterns, an explanation comes to light that makes sense without our having to wave a magic wand.

Shell patterns are so very diverse and complex that the idea of searching for a theory to explain how they're all made seems foolhardy, to say the least. Undeterred, however, that's exactly what some researchers have been trying to do for the last few decades. Just as mathematicians and palaeontologists have set out to describe shell shape, others have done the same for shell patterns.

Their general approach has been to think of these patterns as a form of space-time plot in two dimensions, rather like an inkjet printer. The printer nozzle squirts drops of ink onto a sheet of paper along a straight line and, likewise, the outer rim of a mollusc's mantle secretes pigment into the growing edge of the shell. In both printing and shells, patterns are built up, line by line, as the paper passes through the printer or, much more slowly, as the shell is secreted. Running a finger from the top to the bottom of an inkjet-printed picture, or the pattern on a shell, you're moving through time, from the part laid down first, and hence the oldest, down to the newest. For the printer, digital instructions come down a cable, or through the air, telling it which colours of ink to lay down and when. The question is, what form of instructions do molluscs have to guide them in laying down colours in their shells?

From the start, people tinkering with this question assumed that unlike a computerised printer, molluscs don't carry an image of their complete patterns in their mind which they then break down and reconstruct line by line. Instead, the shell's patterns could be assembled spontaneously at the mantle edge based on a series of relatively simple rules.

In the 1980s, Hans Meinhardt from the Max Planck Institute formulated a computer model that produced astonishing mimics of real shell patterns. Unlike David Raup, Meinhardt didn't spend time thinking about all possible patterns, but was kept busy enough trying to recreate reality. He published a paper in 1987, followed by a book in 1995, *The Algorithmic Beauty of Sea Shells*, which comes with a CD of the MS-DOS program he developed so readers could have a go at decorating their own shells.

Meinhardt's idea was that there could be substances wafting through the mantle that trigger cells to produce pigment. It doesn't so much matter what those substances actually are (they could be hormones or some other form of messenger molecule). What mattered to Meinhardt was their effects; imagine that instead of pumping out drops of coloured ink, a desktop printer produces colourless substances that react with the paper – and each other – in different ways, creating colours and patterns. One of these substances is an activator that switches on pigment production. The activator also triggers the production of more of itself as well as another substance that acts as an inhibitor. Meinhardt predicted that there are antagonistic waves of these activators and inhibitors, chasing each other across the mollusc's mantle edge, stimulating colourful patterns as the shell grows.

At the heart of Meinhardt's model are two differential equations that define how these activator and inhibitor molecules move and interact (and if you like numbers you can find them in his book). By tweaking those equations, he was able to simulate the basic patterns seen in real shells, including all manner of stripes, spots and zigzags.

Stripes parallel to the shell opening are made when pigment production is turned on and off periodically. At first all the pigment cells are stimulated to produce a line of colour, then they are switched off; keep repeating this and stripes unfurl on the growing shell. For bands in the other direction, perpendicular to the shell opening, some pigment cells are switched permanently on and others are permanently off. Meinhardt simulated both of these stripes by altering the relative speeds of the activators and inhibitors in his model.

Diagonal stripes are formed by a process similar to the movement of an epidemic through a human population. A cell loaded with activator can infect neighbouring cells, which after a delay then go on to infect the next-door cells, and so on. This triggers a travelling wave across the array of cells. Interesting things begin to happen when pairs of travelling waves collide. One possibility is they will mutually annihilate each other, drawing a 'V'. Or one wave can annihilate the other, then carry on as a single stripe. Alternatively, they bounce off each other and continue in the opposite direction, drawing an 'X' (although the waves actually cancel each other out, then immediately reignite and continue on their way).

Some travelling waves veer off in different directions while keeping their tails in touch, until suddenly both waves stop in their tracks, creating empty triangles. Waves rushing at each other can also either speed up or slow down, producing spots and teardrops. More involved adjustments to Meinhardt's basic equations lead to more complex patterns, including undulating waves, empty triangles on a dark background and fractal patterns of triangles within triangles, known as the Sierpinski Sieve. All of these shapes and patterns are seen on real shells.

There is, however, one major problem with Meinhardt's ideas: there is no evidence to show that any of this *actually* happens in mollusc shells. No one has ever found a single diffusing substance, no activator or inhibitor, to prove that his ideas are correct. As Meinhardt himself admits in his book, 'Theory can only provide a shopping list of possible mechanisms.'

At around the same time that Meinhardt first published his diffusion model, another research group wrote a paper with an alternative explanation for shell patterns. Bard Ermentrout from the University of Pittsburgh, along with his colleagues George Oster from University of California, Berkeley and John Campbell from UCLA, showed that similar patterns could be created not by unseen substances diffusing around the mantle but via the firing of nerves.

It was Campbell who, in 1982, suggested that the pigment-producing cells in the mollusc's mantle might be stimulated by nerve impulses, just like secretory cells in other animals. The team's model was in effect very similar to Meinhardt's; both simulate a process known as Local Activation with Lateral Inhibition, or LALI. In the 1950s, the great mathematician Alan Turing showed how LALI could work with diffusing molecules, the concept on which Meinhardt based his models. A neural version of this was originally described back in 1865 by Ernst Mach, to explain the optical illusion now known as Mach bands. This occurs when a row of stripes in different shades of the same colour appears to curve inwards from a flat page. This happens because nerves in the back of the eye are activated by the edge of a stripe and will inhibit neighbouring nerves, accentuating the boundary between two stripes. And in a similar way to Meinhardt's diffusing substances, nerve signals can also activate or inhibit the production of pigments and their effect can sweep along, creating travelling waves and various other intricate patterns. Ermentrout and the team implied a very different mechanism to the diffusion model, but made very similar patterns.

The neural and diffusion models had something else in common: Ermentrout, Oster and Campbell also had no proof that their model was correct. Back then no one knew whether nerves do in fact control pigment production in mollusc shells. 'At the time there was no evidence for it, it was just a good idea,' George Oster told me when we chatted on the phone about making shell patterns. After their original paper came out, it would be another 20 years before Ermentrout and Oster published again on shells. When they did, they came closer than anyone ever has to formulating a unified theory that explains not only how seashells get their patterns, but also why they do it.

Decoding the mollusc diaries

If you could listen in on a mollusc's thoughts, the chances are you wouldn't hear anything especially profound because, strictly speaking, they are brainless (unless you're eavesdropping on one of the super-intelligent octopuses, the smartest of all invertebrates). Nevertheless, their simple nervous systems could be responsible for creating the complex decorations that seep across their shells as they grow. The latest computer models that reconstruct shell patterns involve an intriguing new idea: molluscs have the ability to read the patterns on their shells, like the pages of a diary. In this way, patterns become memories etched across their shells.

Shell-making is an expensive business, in terms of getting both raw materials to build them and the energy to lay down new shell. As such, molluscs don't make their shells continually, but in bursts, when they can afford to. Because of the stop-start nature of shell-making, it's vital that molluscs continue construction in the correct orientation, otherwise they'd be all over the place. In their most recent studies, Ermentrout and Oster put forward the idea that shell patterns are a way for molluscs to remind themselves where they left off. This allows them to line up their mantle and continue sculpting their shell in the right places, keeping their intricate shape on track. If this idea is right, then it could be that shell patterns are not quite so useless after all.

Over the last few decades, evidence has been mounting to support the idea that shell-making in molluscs is under neural control. Electron microscopes reveal that mollusc mantles are filled with nerves. These connect back to paired clusters of densely tangled nerves, known as ganglia, that come as close as you will ever get to a general mollusc brain (the ganglia fuse to form a ring through which the oesophagus passes, which means that when a snail swallows, its food goes right through its mind). Nerves stimulate cells in the mantle to secrete new shell layers and, by controlling the amount and direction of material made, different shapes emerge. The mantle also has sensory nerves that seem capable of detecting existing patterns of pigment in the shell. It's possible that each time a mollusc prepares to make more shell, it begins by licking its mantle over the edge of its shell to 'taste' the pattern already laid down. At the same time, nerves in the mantle could also be responsible for switching pigment production on and off.

Based on these ideas and using a revamped version of their 1980s equations, Ermentrout and Oster set out to build a new shell-making program, this time with the help of UC Berkeley grad student Alistair Boettiger. This model not only churned out complex two-dimensional patterns, but it wrapped them around three-dimensional models of shells. For the first time, a realistic mechanism had been formulated for growing shells and decorating them, using a single model.

Whether or not mollusc mantles can actually sense the colours on their shells remains unclear, and it's a major challenge to study real shell-making because it happens so very slowly. A hint that this idea is right, though, comes from the way molluscs repair their shells. As Gary Vermeij knows only too well, in the real, dangerous world it's easy for shells to get whacked, pinched by a crab claw or hurled against a rock. If they survive, molluscs will fix their shells and keep on growing. When a shell becomes damaged or part of it is chipped away, the pattern can get messed up with stripes knocked sideways or stopped in their tracks. But a short way down the line, the pattern usually recovers and continues as before. This suggests that molluscs can detect damage but take a little time to correct themselves. Boettiger's computerised shells do exactly the same thing when inflicted with simulated injuries.

Chaos also features in this latest shell-making program – not that it is all a jumbled mess, but in the mathematical sense. Small changes in initial conditions produce different versions of the same pattern; a little bit of noise here and there makes a real difference. Ermentrout and Oster think this could be why shell patterns in nature can vary substantially between individuals of the same species. Rather than being identical, markings are commonly like human fingerprints, unique to each shell while still sharing similarities in overall pattern.

In 2012, Ermentrout and Oster used their neural model to look at how shell patterns evolve: if they could show that the way patterns change over time isn't completely random, it would support their idea of patterns being useful to molluscs as a way to mark and read their shells. They gathered together a larger team of cell biologists and computer scientists, including Zhenqiang Gong from University of California, Berkeley, who constructed an even fancier computer programme, duplicating 19 species of cone snails that have complex patterns on their shells. The team mapped out a family tree based on the different shell patterns of living species, and used the model to reconstruct what the patterns would have looked like in ancestors further back in the cone snail lineage. They tracked how patterns may have changed over time, as species diverged and split apart. This suggested that some elements of the patterns remained relatively stable over long periods, while others have shifted quickly here and there.

To test the accuracy of their model, the team drew a second family tree, this time using DNA sequences from the cone snails. The match-up between the DNA and pattern-based family trees was striking, far closer than would be expected by chance alone.

All this backs up Ermentrout and Oster's theory that shell patterns aren't frivolous playthings but important registration markers for shell-making that have been subject to the forces of natural selection, and have evolved over time. It may not matter exactly what kind of patterns are made, as long as there is some way for a mollusc to figure out where to put its mantle before continuing to make more shell.

These latest models have undoubtedly taken us a major step closer to understanding how and why molluscs decorate their shells. At the same time, this area of research has cracked open a new window that could have a profound effect on broader reaches of science. The notion that molluscs may leave themselves messages across their shells, allowing them to track the past and make decisions about the future, could give neuroscientists vital clues about how more sophisticated nervous systems work. With this in mind, Ermentrout and

Oster are moving on from gastropods and bivalves to work with brainier molluscs, the cephalopods, and in particular cuttlefish. At least that's what the press release from their 2012 paper said. Both Ermentrout and Oster chuckle when I ask them about this. 'We've talked about it a lot,' Oster says. But the reality is that working with cuttlefish and the stunning patterns they display across their bodies is much more difficult than working with shells. Not only is funding for this sort of research hard to come by, but cuttlefish coloration is much more complex than shell-patterning. Ermentrout and Oster would have to turn their attention from patterns laid down over months to ones triggered in milliseconds. Cuttlefish (and octopuses too) are draped in a mantle that doesn't secrete an external shell, but changes colour to camouflage them or to shout sexy messages to potential mates. These patterns are controlled by a similar network of nerves to those in shell-making molluscs, and, as George Oster points out, 'there's a lot of speculation, but nobody actually knows what the neural circuitry in cuttlefish skin is.'

Nevertheless, I get a strong sense that both of them would love to work with cuttlefish. 'They can flash their colours like gang signs,' Bard Ermentrout tells me. He spends each summer in Woods Hole on Cape Cod in Massachusetts, and clearly enjoys paying a visit to the Oceanography Institution to see the cuttlefish. 'You're not supposed to do this,' he admits, 'but if you put your hand in and touch them, the image of your finger remains on their skin for a few seconds. It's really cool.'

If Ermentrout and Oster can find a way of working with cuttlefish, then perhaps by understanding how their networks of nerves create patterns of 'thoughts' across their skin, it could ultimately help reveal how human brains form memories, deep down where no one can see them.

CHAPTER THREE

Sex, Death and Gems

Forty years ago, in the city of Varna on the Bulgarian coast of the Black Sea, workmen were digging a trench to lay a power cable when they stumbled on something unexpected: human remains – very old human remains – and a great hoard of gold treasure. Archaeologists quickly stepped in and uncovered the rest of a vast necropolis, a prehistoric city of the dead, comprising at least 300 graves that had been dug more than six and a half thousand years ago.

The glittering gold that first caught the workmen's eyes turned out to be part of the oldest haul of buried gold known in Europe. But the gold jewellery and ornaments were not the only treasures left in these graves. In the finest of them all, the resting place of the most powerful man of this ancient community, was a circular bracelet carved from a single seashell that came from far away. It was carried hundreds of miles overland and given to a skilled artisan, who spent many hours carefully polishing and carving it. When it was finished, the shell bracelet was snapped in two, then fixed back together with strips of gold plate hammered with rows of fine dimples.

No one knows exactly why this shell bracelet was broken and then mended. There is no written record from this time, only a series of objects to tell us about these people of the past. Yet there's little doubt that for the person who made it, and the person who was buried with it, the bracelet held great meaning. The shell had perhaps been just as precious as the gold that was used to fix it – maybe even more so.

Just like the molluscs that use their shells to hunt and dig and move, so people have also fashioned shells into all sorts of objects. Some are practical tools. Archaeologists have found shells made into anvils, choppers, knives, fish-hooks and weights for fishing nets. There are shells that lent themselves to particular uses, based on their size and shape, like the bailer shells (of the genus *Melo*) that seafaring cultures have used for centuries to scoop out water from canoes and sailboats. Ground down into powder, shells are added to animal feeds as a source of calcium. The powder can also be combined with ceramics; pottery made a thousand years ago in the Mississippian culture of North America was commonly made stronger by mixing burnt, crushed shells in with the clay.

Besides the usefulness of shells, people have also admired their elegant shapes, dazzling patterns and gleaming iridescence. It's no great surprise that cultures worldwide have used shells to decorate people and places. What is astonishing, though, is how universally shells have come to hold great meaning. Far from being just pretty things to look at, shells have been embraced as powerful emblems of sex and power, of birth and of death.

For millennia, people in distant corners of the globe have placed whole shells in graves alongside the bodies of their loved ones. Even a long way inland, thousands of miles from the sea, piles of shells lie in ancient burial sites. The dead are interred, sometimes clutching shells in their hands or with cowries placed over their eyes (perhaps because the shells themselves look like eyes). The Scythians, a group of ancient Iranian nomads, roamed the central Asian steppes on horseback, yet made burial mounds decorated with cowrie shells. The Seneca people of New York State believed shells placed in the grave could purify decaying flesh and allow the soul entry to the spirit world. They also made masks with shells for eyes, believing that to look through a shell is to gaze back to the beginning of time. The Winnebago tribe of Nebraska considered shells to be the stars of the sea and the apparitions of dead children, women who died in childbirth and men who died in battle; shells were placed inside sacred caves to honour these dead.

One reason, it's thought, that shells have turned up in so many graves is their colour; in many cultures white represents purity and peace, and, accordingly, it is the colour of birth or death. There is also the notion that shells come from an unseen, watery underworld. Empty shells that wash up on beaches are messengers from the deep. Beachcombers pick them up from the strand line, while pondering the hidden realm they came from, or divers bravely visit this dangerous place themselves and return bearing exotic objects.

Around the world, shells are ancient symbols of sexuality, fertility and renewal, perhaps in part because of their shape. So many people have picked up a cowrie shell, turned it over and seen a lengthwise, dark opening like a corrugated smile that reminded them of female genitalia. Even the cowrie's rounded bump is reminiscent of a pregnant belly. Shells are associated with the life-giving properties of water, and they've come to represent the protective womb, a place of conception and the generation of life.

These ideas help explain why so many creation myths tell stories of shells giving birth to gods, humans and sometimes entire worlds. On the island of Nauru in Micronesia, people tell stories of the god Areop-Enap, who found himself trapped inside a clamshell. He groped around in the dark and found two snails, and made them into the sun and the moon; a worm divided the shell into the sky and the earth, and its sweat dripped down and formed the sea. In the Pacific Northwest of North America, the Haida people believe their creator, the trickster Raven, dug up a cockleshell after a flood and opened it to release the men inside. Raven then persuaded the men to have sex with another mollusc, the chiton, and the resulting offspring were women. And if you think Europeans haven't gone in for creation stories involving shells, just take a look at Botticelli's *Birth of Venus*, with the naked goddess perched on a scallop shell. It's stories like these, and plenty of others, that have tempted people to wear shells as jewellery or sewn into their clothes as symbols of good luck and fertility.

The powerful symbolism of shells can also be heard in the call of shell trumpets. The conch in William Golding's *Lord of the Flies* is a symbol of power – in meetings, only the boy holding the conch is allowed to speak – and it is just one in a long line of emblematic shell instruments that resonate through myths, legends and religions into the distant past. Ancient Indian epics tell of heroes who carried conch shells inscribed with their names, which they used to banish demons and avert natural disasters. Samurai warriors used shell trumpets to relay messages to their troops. The lament of a triton shell trumpet accompanies Fijian chiefs to their graves, and conch trumpets are blown in Haiti to call up Agwe, the voodoo water spirit and protector of ships. Shell trumpets have even left their mark in Hollywood. In Ridley Scott's 1979 film *Alien* the sound of a conch trumpet is used in the soundtrack to evoke the desolate atmosphere of the abandoned spacecraft.

Aztec legends tell of the feathered serpent god Quetzalcóatl, who ventured into the underworld to bring back humans after they were wiped out by a great flood. He struck a deal with the lord of the dead, Mictlanteuctli, who agreed to hand over the human bones on condition that Quetzalcóatl played a conch trumpet. The lord of the dead duped him, producing a solid shell that wouldn't play a note. But Quetzalcóatl outsmarted his opponent and summoned worms to chew holes in the shell and bees to fly around inside; the clamouring insects sent out a hollow roar that showed Quetzalcóatl had stuck to his side of the bargain. The lord of the dead had to let the bones go, and humanity was reborn.

As Quetzalcóatl knew, the key to the conch's use as a musical instrument is its hollow chamber. Just like trumpets, trombones, flugelhorns and other brass instruments, shells have a flared opening, also known as the bell. Cut the tip off a large conch or triton shell, press it to your lips and blow. The buzzing from your lips vibrates the column of air inside and resonates along the bell, emitting sound waves that are sculpted in different ways depending on the shape and size of the shell.

The same physics explains why the sound of the sea gets 'trapped' inside large shells. Hold one to your ear and the hollow space acts as a resonating chamber, picking up ambient noises, the wind or the rush of blood through your ears, modifying and amplifying them until they sound (some say) like the swooshing of waves on a beach.

There are countless stories and uses of shells, from fortune telling and board games to magic amulets that ward off the evil eye. You would be hard pressed to find a society anywhere in the world that doesn't have its own interpretation of these natural objects found in rivers and seas and on land. From them all, I have chosen three shells with three stories that, taken together, show how the shell-makers' homes have captured human imagination from the very beginning, and in their gleaming surfaces we will see many facets of human nature reflected back at us.

The oldest gems

Archaeologists and palaeontologists have various ways of looking into the past and piecing together a picture of how things used to be. When it comes to understanding how humans evolved, the bones of our ancestors reveal a lot about what they looked like, what they ate and the diseases they suffered from. And it is in the objects our ancestors left behind, including some remarkable shells, that we find clues about something else: what they were thinking.

Set into a scrubby hillside near the village of Taforalt in north-eastern Morocco is a huge limestone cave called the Grotte des Pigeons. An international team of archaeologists, led by Abdeljalil Bouzouggar from Rabat University in Morocco and Nick Barton from Oxford University, have been excavating the site for more than five years. They have uncovered stone tools and the bones of African hares and wild horses that show ancient people once lived and ate there. From deep down in the cave floor, in the remains of a fireplace, the team dug up a handful of shells that turned out to have been there for a *very* long time.

The shells are dog whelks (known in North America as nassa mud snails) from the genus *Nassarius*. Each is the size of a thumbnail, cream-coloured with a flattened base and twisted tightly into a neat point. Francesco d'Errico and Marian Vanhaeren of the Centre National de la Recherche Scientifique in Paris scrutinised the shells and deciphered their time-worn story. The dog whelks had traces of red ochre rubbed into them; they were pierced with holes and some have microscopic patterns of wear that indicate they had once been strung on a cord. They didn't come from fossil deposits but must have been brought into the hills from the Mediterranean shore, more than 40 kilometres (25 miles) away at the time. The beachcomber who found them had either selected shells that were already punctured or taken intact shells and later on, perhaps back inside the cave by the fireside, carefully pierced a new hole into each one.

Ashy sediments from the cave revealed how long ago the shells had been left there. One technique, optical dating, involves accessing a chemical clock locked inside grains of quartz and feldspar; the clock slowly ticks away and gets reset every time those minerals are exposed to sunlight. Researchers have learned how to read the clock to calculate how long things have been buried in the dark. Bouzouggar, Barton and the team initially concluded that the cave remains were at least 82,000 years old; repeated tests have pushed the date even further back, to between 100,000 and 125,000 years ago. These pierced and painted shells are the world's oldest jewellery.

Stringing shells onto a cord and wearing them as pendants or beads may seem like a simple enough thing to do, but it reflects a fundamental part of being human. Unlike the stone tools that early hominins were making more than three million years ago to kill and butcher animals, shell jewellery has no obvious practical purpose; it is just for decoration.

However, the care and attention that went into collecting these particular shells, carrying them a long way from the sea, smearing them in red pigment and wearing them shows that they must have meant something important to those early people. We don't know what those shells symbolised, but they hint that people had begun to gain a sense of self-awareness and to think in an abstract manner; they were able to express their ideas about the world around them, and their relationships with each other. What's more, these were not the only dog whelk shells being used as beads in prehistoric Africa. Shell beads made from the same species, *Nassarius gibbosulus*, have been found in similar ancient sites in Israel and Algeria, and others from the same genus have been found in a South African cave. Piecing together these findings, it seems that some time more than 100,000 years ago, *Homo sapiens* living at opposite ends of Africa were using dog whelk shells as decorations.

Until these discoveries were made in Morocco, the oldest known symbolic ornaments were perforated animal teeth and shell beads from Europe, dating back no more than 40,000 years. In contrast to the African bead-makers, who used only a couple of shell types, European beads were made from more than 150 species. This suggests that shell beads played different roles in Europe and Africa, and raises the controversial idea that ancient African beads were a close match to more recent use of shells by hunter-gatherers. Rather than being simply personal ornaments, the African beads may have been passed along interlinking exchange systems, or long-distance networks, that crossed the continent and spanned cultural boundaries. When people lived in the Grotte des Pigeons, the climate was going through rapid change, with fluctuating rainfall patterns that would have made life difficult; perhaps the shell beads helped groups of people to reinforce their cultural identity and get through tough times together.

Even if we can't now be sure what those oldest shell beads meant and how they were used, they are a sign that our distant ancestors were thinking in a thoroughly modern way. Tens of thousands of years later, shells and other artefacts were being made that revealed a new meaning and desire among people: the desire to amass wealth and show off their high status. By the time the Bulgarian shell bracelet was made, then broken, human societies were becoming split in a similar way, and not everybody got to wear shell jewels.

Signs of inequality

The discovery of the Varna necropolis, and the hoards of treasures, transformed the view of so-called Old Europe. This was an obscure and often overlooked period in prehistory, dating from long before ancient civilisations emerged in Greece and Rome, and before Egyptians started building pyramids. Around 6200 BC, farmers were migrating north out of Greece and Macedonia into the Balkan foothills, bringing with them domesticated wheat, barley, sheep and cattle. Until the discovery at Varna, it was generally assumed that society back then, in the Copper or Eneolithic Age, was egalitarian, with people living in small, scattered settlements and no sign of a rich elite. Suddenly, archaeologists found themselves contemplating opulent graves and Europe's oldest stash of gold treasure.

Not all the graves were equally adorned, and some were quite sparse, but the most sumptuous – grave 43 – contained the skeleton of a man who died in his forties, who archaeologists think could have been the leader of the Varna community. He was buried in clothes trimmed in gold and carnelian beads, held a gold sceptre, wore gold earrings and gold bracelets, and each knee was capped with a gold disc; he even wore what appears to be a gold penis-sheath. On his upper left arm, above the elbow, he wore the broken shell bracelet fixed with a gold plate. The shell it was made from, a variety known as *Spondylus*, hadn't come from the Black Sea but was brought to Varna from much further away. It was part of a complex, long-distance trade in valuable luxury goods that stretched for thousands of miles across Europe, and was the first of its kind in the world.

There are still many species of *Spondylus* shells living in seas worldwide, stuck fast to rocks down in the depths, many metres beneath the waves. Their common name is thorny oyster, a perfect description for these bivalves with their shaggy coats of spines that encourage seaweeds and other organisms to settle, lending them a cloak of camouflage. The shells themselves are commonly a deep orange, purple or blood red, but in life they are often smothered in encrusting sponges like a colourful, gloopy sneeze.

Most of the ancient *Spondylus* artefacts found across Europe were made from shells collected while the molluscs were still alive. There are few signs of wear and tear that would suggest they spent time rolling in the surf before a beachcomber came along and picked them up. It also seems unlikely that these shells came from fossil deposits. To collect them, people must have found the places where they grew and pulled them from the rocks they clung to. But where did they go to find these shells?

In 1970, when Nick Shackleton and Colin Renfrew analysed the oxygen isotopes in ancient *Spondylus* objects, they found a chemical signature that was etched into the shells while they grew. This revealed their Mediterranean origins, and in particular the warm, clear waters of the Aegean Sea. It was here, in the early part of the Neolithic (around the seventh or sixth millennium BC), that fishermen began gathering *Spondylus* shells. They probably used rakes, dredges and perhaps even tongs from the surface to pluck shells out of the depths; skin divers would swim down and chip the oysters away with knives while holding their breath. Divers and fishermen then passed their shells on to local artisans, who transformed the raw material into all sorts of bright, white ornaments. *Spondylus* beads, buttons, bangles, pendants and belt buckles have been found – mainly in graves – throughout the Balkan Peninsula, in Ukraine, Hungary and Poland, in Germany and westwards into France, where a cylindrical *Spondylus* bead has even been found on the outskirts of Paris.

For these Mediterranean shells to have become so widely dispersed, there must have been a major network of people travelling around Old Europe, meeting each other, exchanging goods and at the same time swapping knowledge and ideas. The popularity of *Spondylus* grew throughout the Copper Age, especially in areas far from the coasts. Then, all of a sudden at the beginning of the Bronze Age, around 3,000 years after they first appeared, *Spondylus* objects vanished from the archaeological record. Either the shells were no longer available, perhaps because the social networks supplying them broke down (there's no indication that the shells had been overfished at that time), or maybe people simply didn't want them any more.

The meaning instilled in all these objects made from Aegean *Spondylus* remains part of what archaeologist Michel Louis Séfériadès described as a 'halo of mysteries'. There is no doubting their value and deep significance, given how many people across such a large area buried their dead with them. Accumulating objects made not just from shells but from gold, copper and other exotic materials seems to have been a sign of high rank or prestige, the preserve of chiefs and revered elders. Many *Spondylus* objects are rubbed and worn in ways that suggest they were used for a long time and passed between people, picking up stories and becoming heirlooms. Remains of a few workshops have been uncovered, further from the Aegean coast, where people reworked and recycled shell artefacts, which must have been a valuable and limited resource. Especially intriguing are the items that were deliberately damaged after they were made.

Archaeologists have uncovered many broken *Spondylus* objects, and at first it was assumed that they were mistakes, evidence of artisans whose hands had slipped. But it soon became obvious that these were no accidents.

One theory is that breaking and burning shell objects was a ritual of conspicuous consumption, a flamboyant way of asserting your status and showing who's boss. It could also have had a more spiritual basis. In 2006, John Chapman and Bisserka Gaydarska, from Durham University, led a team who brought together most of the known *Spondylus* bracelets from the Varna necropolis, more than 200 in total. Like a giant jigsaw puzzle, they tried to work out which pieces fitted together. They found that many, but usually not all, of the parts of a fragmented ring were placed together in a single grave; there were often pieces missing.

It's possible that rings were ceremonially broken at the graveside; some fragments were buried with the deceased, with the rest given to mourning friends and relatives, creating indelible links between the living and the dead. It's also possible that broken rings were used to create and maintain links between living people, who smashed and shared out a ring, carrying the parts of it around, before reuniting them in the grave. Across Old Europe, there are other objects that seem to have been carefully manufactured and then deliberately destroyed, including little clay figurines that were thrown into fires and ritually exploded.

Something else archaeologists have done with the ancient *Spondylus* rings is try them on. Chapman and Gaydarska found that many of the complete bracelets were too small for either of them to slip over their adult hands. But a younger volunteer, a five-and-a-half-year-old boy, could wear most of them (presumably under close supervision) and even fit some bracelets over his feet and onto his ankles. People from Old Europe may have ritually worn *Spondylus* rings from childhood, keeping them in place and soon being unable to take them off again.

As for the bracelet from Varna that was broken and then fixed back together with gold plates, this seems to have been imbued with even greater meaning. Michel Louis Séfériadès thinks it could be evidence of shamanism in Old Europe. He suggests that many things made from *Spondylus* were the ritualistic paraphernalia of shamans, part of a magical tool kit for communing with the spirit world. Maybe the only way for the buried chief to take his jewellery with him into the afterlife was to break it first – to make it imperfect.

Many thousands of years later, on the other side of the globe, parallel trades in *Spondylus* shells emerged, and there too ideas of shamanism flourished. In pre-Columbian times, Mesoamerican and Andean societies placed immense value on these shells, using them in some similar ways to Old Europeans. Archaeologists have traced *Spondylus* all over the region, from Aztec tombs to Mayan iconography and Incan carvings. Starting in around 2600 BC, divers ventured beneath the waves and collected the two species of Pacific *Spondylus* that inhabit the coasts of modern-day Peru and Ecuador. The shells were carved into beads and used as inlays for fine jewellery, often keeping the orange, purple and red colours. Masses of tiny shell beads, known as *chaquira*, were made by the Moche people in northern Peru; a hoard of close to 700,000 *chaquira* was found in a deep tomb in the suburbs of Quito. Beads were often strung together into clothes, including a form of body armour worn by warriors.

As in Europe, shells found in graves reveal how stratified cultures were in this part of the world, with the rich elites accompanied into the afterlife by bounties of oceanic treasures. Unlike in Europe, though, whole shells were often left as grave offerings; nearly 200 enormous *Spondylus* shells, each weighing up to a kilogram, were placed inside a tomb built by the Lambayeque culture in Peru around 1000 AD.

The symbolism of *Spondylus* ran deep, with not only real shells but also ceramic replicas and shell images in murals and sculptures. In the ancient city of Teotihuacan, 30 miles outside Mexico City, plumed serpents carved from basalt swim along the sides of the temple of Quetzalcóatl, weaving between depictions of *Spondylus* shells. There were links to agriculture, with shells offered up to the gods to bring rain and prevent drought.

People also ate *Spondylus* meat, although perhaps not simply as food. Images depicting these shells being held and eaten by deities have prompted some ethnographers to suggest that the shellfish were a source of mind-altering drugs. At certain times of year, warm seas can become stained blood red with blooms of toxic algae. For a time after a 'red tide' has hit, many shellfish become poisonous to humans; the molluscs absorb neurotoxins from the microscopic algae and pass them on to anyone who eats them. Symptoms of

paralytic shellfish poisoning vary; it can make you feel numb
and giddy, and sometimes as if you're flying, but a large dose
can be lethal. There is evidence that shamans in early Andean
societies used various plants and animals, including toads, for
their psychotropic effects. Mary Glowacki from the Florida
Bureau of Archaeological Research thinks they could
have also used poisonous shellfish to help them communicate
with the supernatural. Her theory is that shamans may have
learned to read the tides and predict when a moderate dose
of poisoned shellfish could trigger an out-of-body experience.
Given the way that the human kidneys excrete the toxin, it's
even possible that drinking the urine of someone who ate
infected *Spondylus* would get you high.

Spondylus has been linked with other gruesome practices
in Aztec society. Beneath Teotihuacan's temple of
Quetzalcóatl, 60 human sacrifices were buried with their
hands tied behind their backs. They were dressed in garlands
of *Spondylus* shells, carved to look like human teeth, and
arranged as gaping jaws around their necks.

The complex and occasionally blood-curdling history of
these shells travels into the high peaks of the Andes. In the Inca
Empire, children were led by a procession of priests into the
highest, most sacred mountains, where they were ritually
sacrificed, allowing them entry into the realm of the gods –
supposedly a great honour. At such high altitude, the victims'
bodies have occasionally been preserved by the freezing, dry
conditions; they look as though they have simply fallen asleep.

One of these mummified discoveries was a 12-year-old
girl, who was found in 1996, some 500 years after she died.
She was curled up on a platform facing the rising sun, at the
peak of Sara Sara, a volcano in southern Peru. The team of
high-altitude archaeologists who found her, led by Johan
Reinhard, called her Sarita, 'little Sara'.

Several other sacrificial children were found nearby, along
with a collection of luxury artefacts: miniature human
effigies made from gold and silver, bundles of coca leaves

that were chewed to stave off altitude sickness, and statues of
llamas carved from *Spondylus* shells, unmistakable with their
long ears standing to attention. Most intricate of all these
objects was a male figurine, roughly the size of an Academy
Award Oscar statuette, made from silver and adorned with
fragments of cloth. You can see his finely shaped toes, and
ears pulled into long lobes; his hands are folded across his
chest, and he wears an ornate headdress fashioned from red
Spondylus shell. All of these shell objects had been on a long
journey, 5,000 metres (more than 16,000 feet) up into the
clouds, a very long way from the ocean they came from.

Wind the clock forward a couple of hundred years and
we find people still using shells to gather wealth and status,
but on a scale never seen before, and in a way that combined
ideas both ancient and new. The story of these shells reveals
an even darker side to human nature.

Turning cowries into currency

Shallow coral lagoons in the northern reaches of the Indian
Ocean are home to a small but immensely prolific seashell,
the Money Cowrie. The shells, often three centimetres (one
inch) long, are creamy white and lumpy, sometimes with a
dainty gold line encircling a central hump. In life they are far
more stunning than in death; the shell is covered by a frilly
black and white mantle, intricately patterned like a miniature
zebra.

Throughout most of their lives, cowries inhabit nooks of
coral reefs or the branching fronds of seaweeds, and they
don't travel far. Female cowries lay clutches of eggs and sit
on them, before they hatch into minute larvae. Then, for a
short window of time, her offspring become travellers; the
larvae drift around for a while in the water, riding the
currents and tides, before settling down to live out their
time as ponderous adults. However, after they died, the shells
of many millions of cowries were once taken on long
journeys, journeys with a sorrowful end.

Centuries ago, people in the Maldives began gathering cowries from the warm waters around their islands. They didn't use tiny fishing lines and baited hooks, as one early traveller dubiously reported, but took advantage of the cowries' secretive nature. The easiest way to harvest these shells was to throw coconut palm leaves into the shallows, then leave them there for several months. In that time, cowries would come out of hiding and investigate this new source of food and shelter, taking up lodging among the leaves. All the cowrie-fisher needed to do was pull a palm frond out of the water, give it a good shake and the cowries would drop off. It was then a matter of removing the snails from their homes by burying them in hot sand for a few more months. The end result was a stash of gleaming empty cowries, ready to be sorted and packed into triangular bundles wrapped in coconut fibre cloth. At last, when the monsoon winds began blowing from the south, wooden sailboats were cast off and the cowries began a new journey.

The first port of call was India, where the cowries were exchanged for rice and cloth under the strict control of the Maldivian king. No one else was allowed to take part in the trade. Some of these cowries stayed in India and were used as decorations, amulets and symbols of purity. Indians also used the shells as hard currency, to pay taxes and ferrymen at river crossings. And from possibly as early as the eleventh century, the cowrie trail spread to more distant lands.

Arab merchants took cowries from India on a shadowy overland route across the Sahara. Little is known about these early traders beyond snippets of evidence here and there; some archaeologists believe cowries were traded in Cairo in the Middle Ages, and in the far west of the Arab world, in Mauritania, remains have been found of an abandoned caravan, complete with its cargo of cowries.

Maldivian shells were first traded in West Africa in small quantities as amulets and charms, something that native shell species were already used for. By the fourteenth century, cowries had been adopted as currency. The Money Cowrie doesn't inhabit West Africa, so all the cowries in the region were imported from afar. In the mid-fourteenth century the great Moroccan explorer, Ibn Battuta, wrote the first account of cowries changing hands in the Mali Empire. Back then, shells were used in small transactions in the marketplace, to buy food and other domestic goods, as they have been in many other parts of the world.

Shells are one of the oldest and most widespread forms of hard currency. In New Guinea, people have pierced flakes of pearl shells and threaded them onto strings, measured across the chest in nipple-to-nipple lengths; Native Americans of southern New England made tubular beads, known as wampum, from whelk and quahog shells, which became legal tender when European settlers arrived; and in the Pacific Northwest, from Canada to California, strings of tusk shells (scaphopods) were used as money. In China, the use of cowries as currency goes back thousands of years. The classical Chinese character for money stems from a pictograph of a cowrie, and when demand outstripped supplies of real shells, people made imitations from bone, ceramics and metal. And it could be that the ancient trades in *Spondylus* shells, on opposite sides of the world, also included a form of currency which, some say, is the origin of the word 'spondoolies'.

Shells work well as a form of money for various reasons: they are difficult to fake convincingly; many of them (cowries in particular) are of a consistent size and weight; they are tough and durable; and they feel nice in your hand and are easy to handle. The deep symbolism of shells, and their association with power and status, may also have encouraged their use for important transactions such as marriage dowries.

The trade in shells between the Indian Ocean and West Africa continued on a small scale for several centuries.

It wasn't until European traders came on the scene that a radical shift took place and a whole new commodity emerged that could be purchased with shells, one that would change the course of human history.

Portuguese merchants were the first to figure out the connection between seashells from the Maldives and the markets of West Africa. For a while, they had the trade by sea to themselves but the British and the Dutch soon joined them, and eventually took over. Between 1600 and 1850, the East India Companies of these two great trading powers dominated global shell commerce.

Fleets of ships, known as East Indiamen, sailed first to India, Indonesia and China, where they loaded up with fine goods that were in great demand back in Europe: silks, spices and tea. Before departing again for home, the crews would stop at Indian and Sri Lankan ports to fill their holds with millions of Maldivian cowries. At this point of the trade, the shells were cheap and their main purpose was to act as ballast to keep the ships stable in rough seas throughout their voyages across the Indian Ocean, around the Cape of Good Hope, up the west coast of Africa and back to Europe.

The shells were unloaded into auction houses in Amsterdam and London, where another circle of traders were waiting. They snapped up the shells, repacked them into a second fleet of ships and sailed them back down south.

Some two years after they had been plucked from the Indian Ocean, millions of cowries ended their longest ever journey. In the final stage of a 15,000-mile trip, the shells were lowered over the side of European ships and into small canoes that paddled up the shallow, mangrove-fringed creeks of West Africa. The shells were to be exchanged, not for goods to ship back to Europe, but for human slaves.

European slave traders had discovered that shells were the ideal currency to take to Africa and trade with kings and merchants (ammunition, weapons and other factory-made goods were also exchanged for human lives). Traders turned a handsome profit, importing dirt-cheap shells and exchanging them for slaves.

Prices per human head rose over the years. In the 1680s, a slave cost around 10,000 shells; by the 1770s the price tag hanging around the neck of an adult male slave was more than 150,000 cowries. Once the shells had changed hands, the slaves were shipped across the Atlantic, many of them to work in Caribbean plantations. And so it was that cups of English tea, made from tea leaves packed among Maldivian cowries, were sweetened with sugar grown by the men and women whose lives had been bought with the very same shells.

At the peak of the slave trade, British fleets were importing an average of 40 million cowries into West Africa every year. Throughout the eighteenth century, as Jan Hogendorn and Marion Johnson discuss in detail in their book *The Shell Money of the Slave Trade*, 10 billion shells were shipped across the Indian and Atlantic Oceans.

From the point of view of the molluscs that made all those shells, this is a hugely impressive feat. Enduring such intense exploitation without dwindling is testament to their reproductive prowess and it comes as rather a surprise given that female cowries must spend much of their time brooding eggs, instead of casting their young straight into the big blue, as many of their relatives do. In general, the longer an animal spends tending its offspring, and the fewer young it produces in one go, the more vulnerable the population is to overexploitation by humans.

When the trade in Maldivian cowries collapsed, it was not because supplies of shells had run out. In 1807, the British government passed an Act of Parliament making the slave trade illegal throughout the British Empire, and although trafficking persisted for a time among some colonies, the trade in shell money to West Africa drew quickly to a halt. Humans would never again be swapped for shells on the international market, although for a time slaves were still sold within Africa for shells. But this wasn't the end of the story for the European trade in shell money. A decade later, another new commodity emerged in West Africa, which once again was shipped to Europe in return for shells. Europeans turned their attention from exploiting fellow human beings to exploiting the natural world, and they did so on an even more staggering scale.

It's perhaps strange to think that the global trade in palm oil that is currently responsible for the bulldozing of natural habitats across the tropics has its origins in the nineteenth century. Palm oil lubricated the gears and greased the wheels of the industrial revolution that set the modern world in motion. Factories and homes were lit with palm oil lamps, and workers used palm oil soap to wash off the factory grime.

Back then, most of the world's palm oil was grown in West African plantations, and British traders continued to use Maldivian cowries to buy it. Rather than fading away, the shell trade ramped up a gear, more than doubling previous levels. By 1850, more than 100 million shells were being traded each year. But there was one more crisis ahead for the European shell trade, one from which it would never recover.

In 1845 a German trader, Adolph Jacob Hertz, sailed west across the Indian Ocean after unsuccessfully trying to buy cowries directly from the King of the Maldives. The Maldivian monarchs had always been hostile towards any European merchants who showed up at their islands and Hertz was no exception. On his way back home, he called in to Zanzibar, an island off Africa's east coast, where he discovered an all-too-obvious truth: cowries live all over the place.

On Zanzibar's fine white beaches, Hertz found the Gold Ringer Cowrie. This species is similar to the Money Cowrie,

although slightly larger and with a more prominent golden
circle on its back. Many traders had known of gold ringers
and considered using them, but so far this alternative hadn't
made a dent in the Maldivian cowrie trade, largely because
African merchants refused to accept them. However, the
time was right for Hertz, and his discovery went on to
revolutionise the cowrie trade. He set sail from Zanzibar,
taking with him a few gold ringers and a good idea of
where to find plenty more.

Before long, a trickle of gold ringers began to enter
markets in West Africa. Exactly why merchants finally agreed
to take these alternatives remains unclear. It could have been
the impact of the booming palm oil industry that was
pushing up prices of Money Cowries so that traders
welcomed a cheaper option. These new shells went into
circulation alongside the traditional Money Cowries, and
the trade from East Africa soared.

This time around it was private dealers who dominated
the shell trade, rather than national companies. German and
French fleets transported gold ringers directly from East to
West Africa and in less than 20 years imported 16 billion
cowries, almost as many as the British and Dutch had
throughout the previous century.

Gold ringers flooded into West Africa with a swift and
inevitable consequence. Hyperinflation gripped the trade,
and the value of shell money plummeted. Soon a handful of
cowries was all but worthless. The Maldivian harvest of
Money Cowries had already slumped and, 600 years after
the shell trade began, it finally came to an end.

By the opening decade of the twentieth century, imported
cowries had changed hands as currency for the final time. In
total, more than 30 billion Maldivian cowrie shells ended up
half a world away from where they were hatched and lived.
The nature of shell money means they could not be
withdrawn from circulation or replaced. Some shells were
crushed for limestone and many were built into walls and

floors, as reminders of former wealth. And some people buried hoards of cowries, hoping their riches would once again be worth something. A day that would never come.

As well as all the cowries imported into West Africa as tainted symbols of oppression, the region has plenty of shells of its own. Most of them aren't used for money, though, but for food.

CHAPTER FOUR

Shell Food

Not far from the westernmost point of the African continent, on a cool cloudy afternoon, I stood gazing up at the bare branches of a baobab tree. Its crown, 10 metres above me, looked out over the mangrove forests and the salty, winding creeks of Senegal's Sine-Saloum Delta that flow into the Atlantic Ocean. As all baobabs do, this tree had a gargantuan trunk with folded, blubbery skin. The spongy insides hold a water reservoir that sees it through dry times. Shortly before the rains return each year, the baobab draws on this pool of water and bursts into blooms of dangling white flowers that stink of rotting meat, attracting bat pollinators. Over this tree's long life – at least a few hundred years – it has seen many rains and many bats come and go. And down beneath its roots, this giant, ancient tree has been growing for all these years on a vast pile of seashells. Over centuries, millions of empty shells have been bound together in the soil. I was standing on an island made of shells.

More than 200 shell middens have been found across the delta. The oldest dates back more than 10,000 years, with the largest standing 11 metres (more than 35 feet) high and spreading across an area of 10 hectares (25 acres). Several are burial tumuli, the final resting place for rulers from the kingdoms of Sine and Saloum, which share a distant, entwined history. Other shell mounds contain no human remains but are just the accumulated debris from millions of molluscs that have been eaten by people.

Cockles and oysters have long been a staple food for people living in the Sine-Saloum Delta and they have been the basis for an export trade since the sixteenth century. Mandinka merchants harvested seashells and sold the sun-dried meat far and wide. The piles of shells they left behind are testament to the immensely rich waters that have produced so much food over the millennia.

On one side of the shell island I crunched along a beach made entirely of cream and grey cockleshells. These are West African Bloody Cockles. Their name comes from the bright red haemoglobin pigment they produce (rather than transporting oxygen around the body, as it does in vertebrates, haemoglobin in bloody cockles could have a role in disease resistance). Beyond the beach, a mangrove forest began. The boatman who brought me to the little island steered the narrow wooden pirogue into a green tunnel of these salt-loving trees. I clambered onto the tough mangrove roots to get a crab's-eye view of the world. When the boat engine cut to silence I could hear snapping and popping all around me. It was the sound of oysters, shutting their shells as the tide fell. Known as Mangrove Oysters, they live permanently stuck to these shadowy roots, and twice a day, while exposed to the air, they stay firmly closed, holding a miniature salty ocean inside their shells. The oysters' dry spells are far shorter and more frequent than the baobab's prolonged, yearly

droughts. Listening to all the clops and crackles, I realised I
was surrounded by a vast and noisy seafood feast.

Molluscs have always been an important food for people,
and it's easy to see why. Unlike many more mobile sea
creatures they don't tend to hurry away at great speed, and
they live in shallow waters and between the tides, so even
the clumsiest hunter can easily pick them up. Plus they
come neatly packed into their own containers and cooking
pots. Some varieties taste better than others and while there
can be complications, as we will see, on the whole molluscs
are nutritious and protein-packed.

At the moment, people collectively eat more than 16
million tonnes of molluscs every year, worth approximately
$5 billion, more than £3 billion (the total amount of fish and
other sea life we eat annually is around 130 million tonnes).
Most of those molluscs are bivalves, and most of them don't
come from the wild but are reared in seafood farms, with
more than 70 per cent of them grown in China. The molluscs
we eat tend to be grouped together with various other
shelled marine animals – mainly crabs, lobsters, prawns and
shrimps – and are referred to collectively as shellfish.

In the past, people have had some rather peculiar shellfish
habits. The Romans apparently liked to eat clams that glowed
in the dark. Pliny the Elder wrote about people's mouths
shining like fire, with bright juices dripping over their hands,
down their tunics and onto the floor. Piles of empty mollusc
shells have been found at the site of a Roman bathhouse in
southern England, including bioluminescent angelwing clams
(also known as piddocks). Were these the leftovers from night-
time bathers, feasting on a twinkling midnight snack?

These days, mollusc-eating is usually a much less adventurous
pursuit, especially in Britain. Many people have mixed feelings
about eating food that can sometimes look rather like a sneeze,
or has the dismal texture of a mouthful of rubber bands. And

this is nothing new. In her 1867 book *Edible Molluscs of Great Britain and Ireland (With Recipes for Cooking Them)*, Matilda Sophia Lovell laments the nation's lack of interest in anything other than oysters and cockles, especially when compared to the rest of continental Europe where much more enthusiastic mollusc consumption was going on. Whelks, in particular, have long been one of the UK's unloved edible molluscs. Today, thousands of tonnes of these large marine snails are caught around the British coast each year, and almost all of them end up being exported. Huge demand comes from South Korea where canned whelks, known as *bai-top*, are delicacies. Ever since their local whelk stocks collapsed from overfishing more than a decade ago, Korean importers have been hunting for new sources of this sought-after food, and the European snails from a 'cool and clean sea', as one online supplier puts it, are '100% natural'.

Whelks are very easy to catch. Pots, often made from 20-litre plastic canisters that are punched with holes, are baited with crabmeat and lowered onto the seabed. The whelks catch a whiff of food, amble into the pot and are still there stuffing themselves a few days later when the pot is hauled back up. Fishermen traditionally operate a rotational system, harvesting whelks from one area for a while before moving on, and only returning to the same spot when the population has had a chance to recover. As long as there aren't too many people working the same patch of seabed, this is a low-impact way of fishing. It causes none of the physical damage of dredgers that drag heavy metal machinery across the seabed to catch things like scallops, devastating fragile marine habitats as they go. Certainly, it seems to make sense for the British public to get over their squeamishness and eat more of these gently caught, home-grown whelks, rather than sending them to the other side of the world. But having seen a crate of live whelks on sale at a market in Swansea, Wales, and watched their squirming white feet with black freckles, I realised that I still needed convincing that eating them is a good idea.

An excellent reason to eat molluscs is that some of them are among the most sustainable seafood available. With rampant overfishing stripping the oceans bare, there has never been a more important time to consider carefully the available choices when eating seafood. As a result, various conservation groups release advice on which species are the best options, the ones that come from well-managed fisheries that aren't overfishing stocks or vandalising habitats. And the better options include plenty of molluscs.

Rope-grown mussels are often top of the 'good seafood' lists. The process of producing them is simple and smart. Ropes are suspended in the sea or poles are pushed into the seabed downstream from a population of wild mussels. In the spring, as waters warm up and spawning takes place, wild mussel larvae waft through the sea and some will settle on the ropes and poles. Essentially, this exploits the fact that each adult mussel produces millions of offspring and only a handful from each batch will survive naturally. Even if a mussel farm intercepts thousands of youngsters from the rain of larvae pouring through, it will have virtually no impact on the wild population. The most important thing is to have clean water with strong currents that will sweep in larvae and provide oxygen and food for the growing mussels, while dispersing their droppings that otherwise pile up on the seabed and can cause local problems. Once they have settled, the growing larvae can be transferred to wire rafts or mesh socks suspended from the sea surface, where they are left for 12 to 18 months to reach marketable size. Then the mussel farmers come along and gather in their crop by hand.

Farming mussels grown like this involves none of the villains of bad fishing practices. There's no damaging fishing gear that tears up the seabed. There's no bycatch of other, unwanted species that are thrown back, mangled and dead. A major bonus for growing mussels, and other bivalves, is that they feed themselves, sifting particles of food from the water. Many farmed fish are fed on other fish – usually caught from the wild – in particular salmon, as well as tiger and king prawns (known as shrimp in the US). And mussels, on the whole, are also quite a healthy bunch and don't need to be doused in pharmaceuticals to ward off disease, another common practice in fish farms that use powerful drugs to keep diseases under control among animals kept in often cramped, overstocked conditions.

Oyster farms run along similar lines, although a growing number use larvae that are reared in land-based hatcheries rather than gathered from the sea. Mature oysters are kept in aquarium tanks and encouraged to breed, then their larvae are siphoned off and released into nurseries at sea. As they grow bigger, the young oysters can be laid out on racks, placed in cages or glued to ropes. Like mussels, oysters feed themselves from plankton in seawater but they generally take longer to reach marketable size. The most common species, the Pacific Rock Oyster, will be ready to eat after two or three years. With a helping hand from people, this has become the most successful of all the oyster species. It was originally native to the Pacific coasts of Asia, and farmed for centuries in Japan, before other countries caught on. In the twentieth century, they were transferred to oyster farms across the globe and many populations have established in the wild, from Australia to South Africa, Europe to North America. If ever you shuck and slurp an oyster, wherever you are in the world, the chances are you're eating a Pacific Rock Oyster. Which raises a question that I am often asked: is it cruel to gulp down living oysters?

A similar question can be asked of boiling lobsters and crabs alive, and there is mounting scientific evidence to suggest that crustaceans can and do feel pain. However, this remains largely unstudied in molluscs. Basic biology tells us that molluscs in general, and clams, scallops and mussels in particular, are far simpler creatures than crustaceans. The bivalves' lack of brain means they probably have only a limited capacity to sense and respond to the world around them. Among the shelled molluscs we eat, it's the super-intelligent octopuses and squid, plus the itinerant snails, that have heightened senses of perception and are more likely to be able to feel pain – a good excuse, if you want one, to pass on the calamari and escargots.

By contrast, there are some vegans who consider oysters and mussels to be so plant-like that eating them isn't a problem. In reprints of his book *Animal Liberation*, vegan advocate Peter Singer keeps changing his mind about whether it's OK to eat oysters. It seems there is no concrete proof either way on whether they do or don't feel pain. But certainly, I think there are fewer questions to be asked about farming and eating bivalves than there are about rearing mammals and birds in horrific factory farm conditions. However, ethical issues aside, there are a few other reasons why it's not always a good idea to eat molluscs. For starters, eating shellfish comes with the risk of catching a dose of food poisoning, and occasionally something much worse.

The whole shucking truth

On the shores of Chichester Harbour in southern England, the small town of Emsworth was once home to one of the longest-running oyster fisheries in world. Records show that people were eating Emsworth oysters as far back as 1307. The fishery thrived throughout the nineteenth century, but came to a sudden halt in 1902, when tragedy struck. Two great banquets were held in the nearby cities of Winchester and Southampton, where guests were served oysters from Emsworth. In the days that followed, 63 people fell ill and four of them died, including Dr William Stephens, the Dean of Winchester Cathedral. They had all developed typhoid, caught from eating oysters contaminated with human sewage. It turned out that recently laid sewers and drainpipes were pouring effluent into the harbour, directly over the oyster beds. When this unpalatable reality was brought to light, the Emsworth oyster fishery was immediately abandoned, and hundreds of people lost their jobs.

The fact that molluscs – in particular bivalves – can sometimes be very bad to eat is not the molluscs' fault but ours, for polluting the seas they live in. Even in wealthier countries, where most human waste is now collected and treated, sewage still leaks into the sea, especially following heavy rains when sewers become overloaded. Other effluent comes from farm manure that runs off land, into groundwater and out to rivers and coasts. And whenever there are disease-causing bacteria and viruses floating around, it doesn't take long for bivalves to pick them up; oysters can filter around 100 litres of water every day, a large bathtub-full. The molluscs themselves may not suffer from their noxious load, but they will pass it on to people who eat them. If you're very unlucky you could catch norovirus (also known as the winter vomiting bug), *E. coli*, listeria or salmonella from sewage-infected bivalves. The biggest outbreak of shellfish infections on record was in Shanghai in 1988 when almost 300,000 people contracted hepatitis A from eating clams.

There is a suite of other dangerous conditions people can catch from bivalves. Self-explanatory names describe the key symptoms of various diseases: there is paralytic shellfish poisoning, amnesic shellfish poisoning and diarrhoeal shellfish poisoning, along with neurotoxic shellfish poisoning and the so-called 'possible estuary-associated syndrome'. For vulnerable people, and in high doses, these illnesses can be deadly. The problems stem, once again, from bivalves' habit of filtering seawater for food. A major part of their diet is phytoplankton, the plant-like microbes that harness the sun's energy on a colossal scale. Among thousands of phytoplankton species there are around 80 that can become extremely virulent, including some dinoflagellates and diatoms. They produce strangely named noxious compounds like yessotoxin, saxitoxin and domoic acid, which between them cause the various shellfish poisoning syndromes.

These toxic phytoplankton will sometimes proliferate, triggering phenomena formerly known as red tides, though scientists now prefer the term 'harmful algal blooms' because they can turn water purple or green or dingy brown. When bivalves find themselves in the middle of a harmful algal bloom, they filter plankton from the water, along with the toxins. Again, the bivalves themselves don't suffer, but the toxins build up in their tissues to levels that are dangerous for anyone or anything that eats them (including, perhaps, those ancient Andean shamans, who may have eaten them to commune with the spirit world). Harmful algal blooms can happen quite naturally without any influence from people. Recently, palaeontologists uncovered a nine-million-year-old graveyard of 40 or more whales in the Atacama Desert in northern Chile that were thought to have died after eating fish contaminated with toxic plankton.

The really bad news is that harmful algal blooms seem to be on the rise. Several decades ago, they were only seen on a few coastlines, but outbreaks now appear worldwide. At the same time, shellfish poisonings are more common, and molluscs are generally becoming more dangerous to eat. In the 1980s, there were around 2,000 reported cases of shellfish poisoning each year. More recently, estimates have shot past the 60,000 mark. It may be that people have become more aware of the issues and report a greater proportion of cases than before. There are also a lot more people in the world, and more of them than ever are eating shellfish. In China alone, clam consumption has increased 400-fold in the last 30 years. However, the most likely explanation is the fact that people are poisoning the seas – although this time, indirectly.

The exact triggers for harmful algal blooms are still being studied but one important factor is well established: nutrients. Wherever nitrates and phosphates wash into seas and lakes it increases the chances of a harmful bloom forming because phytoplankton absorb those nutrients and grow, just like plants on land when fertiliser is added to the soil. The extra nutrients can very quickly kindle a lot more plankton.

The rise of artificial fertilisers and industrial-scale farming have played a big part in nutrient pollution. Since the industrial revolution, average phosphate levels in coastal waters have tripled, and nitrate levels have increased even further. Household cleaning products are also implicated. Environmentalists are campaigning for these nutrients to be banned, and in recent years the EU and US have set strict limits on phosphate levels in domestic laundry powders and dishwasher detergents (people living in hard-water areas will just have to make do with glassware that doesn't gleam quite so brightly). The massive growth in fish farms in recent decades, especially for salmon, is also contributing to the flood of nutrients into the seas, from fish faeces and uneaten fish food.

As well as encouraging plankton blooms, the torrent of nutrients pouring into the ocean triggers a second ecological disaster. When the blooms come to an end, usually after several days, weeks or months, the dead plankton sink to the bottom where bacteria break them down. This uses up oxygen from the water and creates so-called dead zones where few aquatic species can survive. Since the 1960s, the number of dead zones worldwide has doubled every decade. One of the largest and most persistent is in the Gulf of Mexico, fringing the US states of Texas and Louisiana, which is caused largely by the polluted waters of the Mississippi River. In 2014 it covered 13,000 square kilometres (5,000 square miles), an area roughly the size of Connecticut or East Anglia. And as climate change warms up the oceans' nutrient-rich soup, the extent and duration of harmful blooms and dead zones will only get worse.

Because of all these threats, many countries have introduced checks and balances to ensure shellfish is safe to eat. Early warning systems forecast and detect the onset of harmful algal blooms and, when they do strike, any nearby fisheries and fish farms will be closed until all risk of contamination has passed. Regular tests are carried out in many countries to check on levels of bacteria and toxins in shellfish. And while raw sewage may not always be pumped into the sea as it was in days gone by, coastal pollution still remains an issue.

Across Europe, shellfish beds are assigned to strict classifications according to the levels of faecal coliform bacteria they contain. Grade A molluscs can be eaten straight from the sea (they have fewer than 230 *E. coli* cells per 100g of flesh). Meanwhile Grade B molluscs, with higher coliform counts (up to 4,600 *E. coli* per 100g), must be put through a process of purification (or depuration) before they're eaten. Following the Emsworth oyster poisonings and various other typhoid outbreaks in Europe and the US, methods were developed for depurating bivalves. Now a well-established technique, it generally involves keeping live bivalves for 42 hours in tanks of fast-flowing clean water, often blitzed with UV light, to remove contaminants from their tissues. Some oysters are put through a depuration process even when they come from Grade A beds, just to make sure. There is also a Grade C (for shellfish with up to 60,000 *E. coli* per 100g); these have to be moved to cleaner coastal waters before anyone is allowed to consider eating them. Molluscs with even higher coliform counts are strictly off limits.

Today, the majority of molluscs that make it to market – at least in developed countries – are fine to eat but only because we've had to invent ways of protecting ourselves from the pollutants we pour into the natural world. When it comes to eating molluscs, the other major issue is that some species are rather too delicious for their own good. A long time before we came up with ways of farming them, humans already had a terrible track record of taking too many molluscs from the sea.

Who ate all the clams?
The very earliest known case of any wild species being driven to the brink of extinction by people was a giant clam, around 125,000 years ago. Giant clams are the largest living seashells on the planet. They can grow to well over a metre (three feet)

in length, and live for longer than a century. I saw a living giant clam for the first time many years ago on Australia's Great Barrier Reef and was amazed at just how huge it was. I smiled down at it and it seemed to grin back with its colourful, corrugated lips (their fleshy mantles acquire their bright colours from photosynthetic microbes called zooxanthellae living inside them, similar to those that live inside many corals). The clam sensed my shadow passing over it with hundreds of tiny eyes and hesitantly withdrew its mantle and closed its twinned shells. Their reputation as dangerous man-traps is utter nonsense with no record of anyone ever getting a part of themselves stuck inside one of these enormous bivalves. As with many legendary beasts, giant clams have far more to worry about from us than we do from them.

A few years ago, while exploring the warm clear waters of the Red Sea, a team of divers found a species of giant clam that no one had seen before. When he first caught sight of the deeply crimped shell, Claudio Richter from the Alfred Wegener Institute in Germany suspected this was something different to the seven known giant clam species. Further physical analysis and DNA tests confirmed the species was new to science and the team named it *Tridacna costata* (from the Latin word *costatus*, meaning ribbed). Scouring the reefs across the Gulf of Aqaba and the northern Red Sea, the divers found only a smattering of living *Tridacna costata*. To work out whether this has always been the case, the team also hunted for giant clams on land, in the sandy deserts fringing the Red Sea, in fossil reefs that flourished when sea levels were much higher. They saw that *Tridacna costata* was much more common 125,000 years ago, making up more than 80 per cent of the giant clams in the region. Today, they make up less than one per cent of the living clam community. Over the millennia these giant clams have also become distinctly less giant. They have shrunk in size to their present-day dimensions of 30 centimetres, or roughly a foot across. In the past, the clams would have weighed at least 20 times more than they do today.

The most likely explanation for the drastic decline in the clams' stature and abundance is overfishing by humans. When people hunt, they almost always take the biggest animals first and a decline in average body size in a wild population is a good indication that humans have come along and helped themselves. In the case of the giant clams, those early people also left behind their fishing tools. In fossil reefs further along the Red Sea coast, in Eritrea, archaeologists have found palaeolithic stone tools that could have been oyster shuckers, left behind by people who waded out to gather clams and oysters. These gastronomical findings are reshaping our understanding of human migrations, providing new evidence that a coastal route out of Africa may have been important. And the loss of giant clams from the Red Sea, all those thousands of years ago, was just a taste of things to come.

The story of people overconsuming molluscs repeats itself again and again. Huge piles of empty shells show how abundant queen conch used to be across the Caribbean; now they are rare, and continue in their decline despite international efforts to protect them. In kelp forests of the Californian coast, divers have plunged further and further beneath the waves to find valuable abalone. The white and black species are now endangered, while the red and green are heading the same way. We are working our way into the depths and through the colours.

Perhaps the most famous mollusc stocks to collapse were the New York oysters that used to be pulled by the million from the Hudson River. Mark Kurlansky tells their story in *The Big Oyster,* of times when shellfish were eaten in Manhattan only a few blocks from where they grew. As stocks close to New York diminished, fisheries swept along the coast, leaving behind a trail of destruction. The same thing happened on America's western seaboard and in Australia, where similar short-lived fisheries fed demand for oysters in San Francisco and Sydney.

Following all these declines, most of the molluscs we eat today are farmed, but there are a few places in the world where wild oysters still thrive. And it is there that people are trying very hard not to let history repeat itself.

Guardians of the oyster forest

A short way south along the coast from the baobab tree and its shell-island home, I came across more heaps of empty shells. I had crossed the wide mouth of the River Gambia on board a crowded and rusty ferry that crawled slower than walking pace towards Banjul. The Gambia's capital sits on an island where the river meets the Atlantic Ocean. The rest of the country lies to the east, impossibly long and narrow, like a finger poking into Senegal. I continued my journey by taxi, crossing the bridge that links Banjul to the mainland, and by the side of the road I spied a series of silvery grey mounds and a queue of cars pulled over on the hard shoulder. This is where Gambians go to buy oysters.

In The Gambia, as elsewhere in the world, oysters are a delicacy. Gambian oysters also happen to be some of the cheapest you can buy. A bag of smoked oysters, scooped up in an empty tin can, will set you back 25 Gambian dalasis – less than 40 pence – and they come from an extraordinary place. Right on Banjul's doorstep is a swathe of rich, green mangroves. The Tanbi Wetland National Park covers an area slightly smaller than the island of Manhattan. Living in small settlements scattered along the fringes of this aquatic forest are women who venture out and gather oysters from among the mangrove roots. Many of them are the sole breadwinners in their families; the men are either lazy or long gone. I was planning to meet the all-female oyster harvesters and the woman who has been helping them to help themselves while at the same time protecting The Gambia's fragile wetlands.

Fatou Janha, known respectfully as Auntie Fatou, was born and raised in The Gambia but spent many years moving around the world with her diplomat husband; returning home later in life, she decided one day to stop and talk to the oyster sellers.

Since she was a little girl, Fatou had seen women selling oysters by the roadside on the way into Banjul. 'It suddenly occurred to me that these people need help,' Fatou told me, as we sat in her office near the Old Jeshwang market with the voices of songbirds drifting in through the open windows. 'But at first they didn't understand why I should be interested in them. People have ignored them for a long time. As I always say, people buy oysters but they don't look behind the oysters.'

On the day she stopped to talk to them, the oyster harvesters told Fatou about their lives and the troubles they faced making ends meet. She left her number and told them to call if there was any way she could help. She waited, and a few weeks later her phone rang.

When Fatou originally came back to The Gambia she set up a boutique, making and selling clothes. But for many years now, she has been pouring her energies into the TRY Oyster Women's Association, the community project that grew from that first meeting and from Fatou's vision that the women of Tanbi should be allowed to *try* to improve their lives.

Back in 2007 when TRY was founded, there was no denying that Tanbi's natural resources were being pushed too far. Since the 1960s the local population has been growing, with people migrating in from neighbouring Senegal and Guinea-Bissau. Many of them are Jola, an ethnic group that has inhabited these coasts for centuries, in particular the troubled Casamance region of southern Senegal. For a long time they have been living and working in Tanbi. They have always supported themselves and their families by harvesting shellfish from these rich waters but, over the last decade or so, declining catches have forced them to roam deeper into the mangrove forests. The few shellfish they found were all rather scrawny and small. By the time Fatou came along, the oyster gatherers were finding it very difficult to make a living from the forest.

The simple but powerful thing Fatou did was to bring the oyster harvesters together and give them a unified voice. In the beginning, TRY had 40 members from one village. Fatou helped them to get a bank account, raised some funds and set up a micro-finance scheme so the women could start other small businesses and make money during the closed rainy season. She organised classes to teach the women and their daughters to bake and make handicrafts, provided healthcare advice, and encouraged them to start saving for the first time; in particular, she wanted the women to be able to afford school fees and allow their daughters to finish their education. Word soon spread, and now TRY has more than 500 members from 15 villages across Tanbi. Women who a few years previously were strangers have now became close friends and co-workers.

Fatou suggested we pay a visit to the mangroves to see some of the women at work, so we rented a small boat and slowly motored along the tributaries, known as *bolongs*, that percolate through the Tanbi wetlands and divide the forest into a mosaic of islands. Salt-encrusted leaves of mangrove trees rose up around us with their stilt roots looping and dipping into the water. A Malachite Kingfisher flashed past, a dart of electric blue, and a gaggle of pelicans rested in high branches, preening their feathers with enormous beaks. Around 360 bird species are known to inhabit the wetlands, including many global migrants, and birdwatchers fly in from across the world to see them. There are other forest denizens that I didn't see, but it was good to know that Red Colobus monkeys were perched somewhere in the dense thickets, there were short-clawed otters padding through the undergrowth in search of crabs, and perhaps even a manatee was hiding submerged somewhere nearby in the murky waters.

As we chugged through the mangroves, Fatou told me more about oyster harvesting and selling. The women spend hours shucking the oysters, sometimes with the help of younger men in the villages, then they roast and smoke them.

Long-term water monitoring is underway to see if it might be possible to eat them safely uncooked. Fatou eventually hopes to see the oysters on sale in local hotels and restaurants. There are two main types of holidaymakers who flock to The Gambia – wildlife seekers and cheap sun, sea and sand seekers – and hopefully they could both be persuaded to try the local seafood.

Gathering oysters is tough, physical work, but the women much prefer it to being housemaids, the only alternative they see for earning money. The oysters give them independence and a sense of identity; the women now belong to a close-knit sisterhood. Fatou explains that they are some of the poorest people in the country, living in marginalised communities that other Gambians know very little about.

'I want these women to be recognised by society and respected,' she told me. She is incensed that so many Gambians enjoy eating oysters but pay so little attention to where they come from or who collects them. In the short time that I spent with her, I had already seen how Fatou's straight talking and irresistible energy inspires the women and keeps TRY going. Of course, she insisted that it is the women themselves who are strong.

We turned off the boat's engine, and Fatou called out in a high-pitched whoop. Seconds later we heard a reply; this is how the women communicate and locate each other while they're working in the dense forest. Down a side creek, two of them had pulled up their wooden dugout canoe and were busily gathering oysters.

It was early May and the oyster season was in full swing. Until a few years ago, oyster harvesters would leave the mangroves in June, at the start of the rainy season, and return again each December; now they don't come back until March, to give the oysters more of a chance to grow. Being tropical species, mangrove oysters grow much faster than their cool-water cousins, and a couple of months makes a big difference. Within a year of extending the closed season, harvesters were already finding larger oysters, which they could sell for higher prices. An added benefit is the fact that larger molluscs will leave behind more offspring for the next generation.

One of TRY's most pioneering achievements has been an agreement giving them exclusive rights to work in Tanbi. The women of TRY, along with their advisory committee, can now decide who is allowed to collect oysters and issue fines to people who break the rules. This is the first time a group of women in Africa has been granted ownership of an important natural resource, forming the basis of their livelihoods. The wetlands are no longer a free-for-all.

Each village has been put in charge of its own community *bolong*, and there are communal areas where all TRY members can work. In addition to the extended closed season, parts of Tanbi are now set aside on rotation and left alone for much longer to give oyster stocks an even better chance of recovering and replenishing nearby areas. Anyone caught illegally cutting firewood or gathering too many oysters from the wrong place or at the wrong time of year faces a hefty penalty. The agreement also covers West African Bloody Cockles that the women gather from the riverbed. Taking undersized cockles is another finable offence.

None of this would have been possible until the oyster gatherers joined forces under the banner of TRY. The agreement was based on a complex, multi-agency co-management plan that involved departments of forestry and fisheries plus many others. This sort of negotiation is arduous and needed the women to join forces. If they were still individual people working alone, who didn't know and talk to each other, an agreement would never have been reached. And the women haven't simply been handed over the rights to the wetlands to do with as they please: they are committed to looking after them. They are now the official custodians of Tanbi.

The tide was falling and I climbed out of the boat, intending to make my way across the shore to where the women were working, but straight away I got stuck. The mud was up to my knees and sucking at my toes, so I stood where I was, windmilling my arms, trying my best not to make the situation worse. One of the women saw me in trouble and came to my rescue, effortlessly plucking me free and leading me over to higher, firmer ground. I thanked her, in my one bit of Wolof, and she smiled back and continued with her work. She pulled two pairs of socks over her hands to protect her from the sharp shells and using a small knife she nimbly chipped oysters from the exposed mangrove roots. Thick crusts of oysters make the roots look like they've been dipped in lumpy cement porridge.

In the past, some harvesters used machetes to chop away whole roots covered in oysters, big and small; this was wiping out juvenile oysters and damaging the forest itself. Now, as well as being much more selective and careful about taking only individual oysters of the right size, the women of TRY are also trialling an aquaculture technique similar to the one used for mussels, hanging ropes to catch young oysters from the water.

With a basket full, one of the women picked her way across to the water's edge and tipped the oysters into the canoe. I followed and once again got firmly stuck in the mud. This was all getting rather embarrassing and I noticed for the first time that the woman who kept stopping work to help me out, with her incredibly strong, reassuring grip, was at least six months pregnant.

A few days after our trip to the mangroves I got to see another side of Gambian oyster harvesting. Every year, Fatou organises an oyster festival. The idea is to raise money, raise the profile of Gambian oysters and at the same time give the members of TRY a chance to celebrate. Next to the roadside where the women shuck, smoke and sell oysters, set back in a grove of baobab trees, a sandy arena was laid out for the festivities. I arrived just as the women started to parade in. The members of each village were carrying a banner announcing who they were and wearing outfits to match. The costumes on display were all stunning. Some villagers had dresses made from vibrant wax prints trimmed with ruffles and lace; others wore tie-dyed skirts and crisp white shirts, with strings of multi-coloured beads strung around their necks and across their shoulders. Everybody had immaculate hairdos, braided in neat rows and decorated with bright clips, or they wore colourful headscarves matching their dresses, tied into elegant bows and knots. As they promenaded around the arena, the women began to dance and sing, and they wouldn't stop again for another two days.

The band was a troop of tireless young men, four drumming and one with a beaten-up saxophone that he blew as tunefully and incessantly as possible. There was a sound system with noisy speakers and a single microphone for people to sing into, which they were doing without a hint of bashfulness.

Everybody danced, from teenagers to grandmothers, standing in a circle and taking turns to come forward and perform for the crowd, who sang and clapped and cheered. The rhythmical drummers were accompanied by a chorus of whistles, which the women wore on colourful chains around their necks, giving the event the feel of an early nineties rave. This was the first oyster festival I'd been to, and somehow I imagine there are few in the world quite like this one.

The music and dancing paused briefly while speeches were made, mainly for the benefit of attending patrons and dignitaries who sat demurely watching proceedings from the shady marquees. Then the boisterous celebrations continued with a highly unusual performance: the members of TRY took part in a wrestling tournament.

Wrestling is a hugely popular sport in West Africa, but it is normally the preserve of boys and men. Every morning and evening along beaches in Senegal and The Gambia, young men congregate to practise their wrestling moves.

Professional wrestlers are celebrities paid as much as international soccer stars, and contests can draw enormous crowds. It was Fatou's idea to let the women have a go.

'The Jola tribe are known for wrestling, so why can't women wrestle?' she said. Girls wrestle with their siblings at home for fun, she explained, so why not put on a competition for the women of TRY? It was the first time anyone had done something like this, so she could only guess what the response would be. And as it turned out, like most things Fatou sets her mind to, the wrestling at the oyster festival was a runaway success.

Pairs of women and girls stepped up to the sandy ring, wearing wrestling loincloths over their colourful outfits. At the start of each round, the women performed taunting, stompy dances to psych their opponent. Then they locked arms and heads and tried to grapple each other to the ground, all the time marshalled by a referee. Occasionally one wrestler would successfully grab her rival between the legs and launch her into the air, and the crowd went wild. The victor was hoisted on someone's shoulders and processed around the arena, sometimes the loser as well, and mostly I couldn't tell which was which. None of that seemed to matter.

While all that was going on, I watched from the sidelines, happily tucking into platefuls of oysters. The oysters from the mangroves looked and tasted more like mussels than oysters to me; they are smaller than Pacific Rock Oysters, although the two do belong in the same genus, *Crassostrea*. I was busy trying to decide which recipe was the most delicious – oyster spring rolls or the zingy, mustardy oyster *yassa*, a traditional West African dish – when my feasting was interrupted.

Fatou decided it was my turn to enter the ring. I pleaded that I didn't know the rules (and so far I hadn't been able to fathom them out from watching the contest), but she was having none of it. Thankfully, though, she took pity on me and rather than pairing me up with one of the formidable

women from TRY or even one of their athletic teenage
daughters, Fatou instead picked another clueless European
visitor from the crowd.

Our war dances were a tame imitation of everybody else
who had come before us. Then for a few minutes, cheered
on by whoops from the crowd, we pushed and jostled each
other. Rather too soon it became all too obvious that I
hadn't spent enough time wrestling with my sisters as a
child. My adversary pulled a fast one on me, slipping her leg
behind mine and flipping me neatly over on my back. As I
lay in the sand, gazing up at the hazy sun and with an
excited throng of oyster pickers rushing towards me, I
decided it was time to make a hasty retreat back to the food
tent to continue the important business of oyster-tasting.

Deciding which varieties of seafood are better and which are
worse to eat is not a straightforward matter, either for the
environment or for the people involved. It depends on which
species you're eating, where it came from, how it was caught
and who caught it. Dreadful stories have recently emerged
about people in Thailand being forced to work in harrowing
conditions for no pay on fishing boats that dredge up 'trash
fish'. These small, infant, inedible species are scraped up from
the seas, devastating ecosystems, all so they can be ground
down into fishmeal and fish oils and fed to the farmed
prawns, shrimp and fish that are sold in Europe and the US.
Eating Gambian oysters, though, I felt reassured that this
food hadn't come at a great cost to humans or the natural
world. While the women of TRY carried on wrestling,
singing and dancing, it all suddenly seemed to me to be quite
simple: protect crucial habitats (like mangrove forests), don't
eat the oldest, slowest growing species (like giant clams) and
make sure fishers get a decent wage. Sadly, though, simple
truths like this seem to be the exception when they should
be the norm for all the seafood we eat.

CHAPTER FIVE

A Mollusc Called Home

The next time I encounter oysters I can't actually see them, but I know they are there. Standing at the end of Mumbles pier, I look down onto the cloudy, sage-green water, ruffled by westerly winds that are scudding across Swansea Bay and making today a bad day for going out on boats. On the other side of the bay are the looming towers of the steelworks at Port Talbot, a sight often compared to an industrial version of Mordor. Over here on this western side things feel much friendlier.

If the weather had been kinder I would have joined fisheries scientist Andy Woolmer aboard his faithful workboat *Triton*, and helped him hunt for oysters. We would have dragged a small dredge along the seabed, four or five metres beneath the waves, and pulled it up now and then to see what was down there. If the water had been clear, we would have lowered down a video camera to see what we could see.

Instead we walk out across the water. The Mumbles pier was built more than 100 years ago, and is now in the process of being restored; at one end, next to the beach, there's a noisy amusement arcade, and a shiny new lifeboat is stationed at the other end. We look out at the twin bumps of the Mumbles islands (some say their peculiar name originated in their feminine curves. *What a lovely pair of mumbles…?*). Andy points out the orange buoys that bob at the surface, marking out the place where a few years ago he found a derelict oyster bed.

In 1898, when the pier was built, an oyster fishery was still flourishing in Oystermouth, the town that overlooks the Mumbles. It employed hundreds of people, and traded millions of oysters across the British Isles. Ghosts of the fishery haunt the coast today. There are tumbled wrecks of oyster boats and the low walls of enclosures known as perches are still laid out across the sand where oystermen would store their catches for a few days, toughening them up and getting them used to shutting their shells and staying alive in dry air before their long rail journey to London. Oyster bars and stalls used to be a common sight in the town, and local pubs served up carpet-bags — steaks stuffed with oysters — washed down with pints of thick, dark stout. Back then, oysters were a food for everyone, not just the well-to-do.

Until the 1920s, the new oyster season was welcomed each year with a great celebration for the whole town. Using the piles of empty shells from the oyster-processing plant, children constructed shell grottos, like calcium carbonate igloos, and lit candles inside to make them sparkle. People strolling along the seafront would offer pennies to the grotto-makers.

This part of South Wales has a long history of oyster-fishing. More than 300 years ago, the oyster beds at Oystermouth were described as the most prolific in Britain. It is even thought that the Romans gathered oysters from these waters during their occupation of the region from the first century AD; there are the remains of a Roman villa in the town, and recently a midden of oyster shells has been uncovered on Mumbles Hill, overlooking the sea, but this has yet to be dated. In Oystermouth today, however, there are no oystermen, no oyster stalls, no carpet bags and no shell grottos.

It was Andy Woolmer's idea to try to restore the oyster fishery at the Mumbles. If it works, he will bring back not just a lost fishery, but a lost ecosystem as well.

In a similar way to individual trees that grow into a forest or the corals that form the structures of a tropical reef, oysters are also wild architects. Around the world, various oyster species form great gatherings and create beds, banks and reefs that shelter a myriad of other living things. Sedentary species find nooks to nestle in; grazers find green, red and brown seaweeds to chew on; scavengers never go hungry. These assemblages of species, interconnected and relying on each other in so many ways, are what ecosystems are all about. And it was in fact a study of oysters that originally paved the way for the modern science of ecology.

When Karl Möbius explored the Bay of Kiel oyster beds off the German coast, it was primarily with economics in mind. In the 1860s he was tasked by the Prussian government with finding ways to boost harvests of the sought-after and lucrative shellfish. Möbius concluded that farming was not an option for oysters in this part of the Baltic Sea, but from his research a bigger, more important idea came to light.

Getting well acquainted with the great piles of oysters, a mingling of living and dead shells, Möbius was struck by the variety of other creatures that live among them, the fish, crabs, worms and starfish. He was convinced that oyster beds were far richer in life than anywhere else on the seabed, and in 1883 coined the word *biocönosis* to describe these living communities.

His revolutionary idea was to consider not just single species, one at a time, but the teeming throng of interacting life. It would be another 50 years before the term 'ecosystem' arose, taking this concept one step further to embrace both Möbius' living *biocönosis* and the non-living physical environment, the seawater, rain, soil and so on. At the same time that Möbius was laying down some of the foundations of ecology, the oyster beds and their self-made ecosystems, which kick-started his ideas, were being swiftly dismantled.

European coasts were once fringed with oyster beds. It's impossible to know for sure exactly how extensive they used to be, but there are clues. A colourful drawing in the 1883 book *Piscatorial Atlas* shows the distribution of the Native Oyster, *Ostrea edulis*, in Europe. This was based on data gathered by the book's author, Ole Theodor Olsen, who spent years travelling around, talking to fishermen and asking them about the seas and the things they found there. Olsen's map shows that the French, British, German and Dutch coasts were crammed with oysters, as was the English Channel, the Waddenzee and a huge southerly patch of the North Sea.

The immense scale of European oyster habitat was matched, and in time exceeded, by the scale of their exploitation. During the nineteenth-century heyday of oyster fishing, it was said that three men in a small sailboat could easily dredge up 3,000 oysters in a couple of hours. By the middle of the century, half a billion oysters were passing through Billingsgate Market in London each year.

Nevertheless, as with most natural bonanzas, the riches were not to last. By the beginning of the twentieth century, a feast that had been going on since Roman times was shuddering to a halt. New railways had boosted the capacity to meet demand from hungry inland markets; fishing boats became larger and more powerful; dredging gear grew bigger, heavier and more effective at scraping things up from the seabed. To make matters worse, the seas were for the first time becoming seriously polluted by the outpourings of factories and mines as the industrial revolution gathered pace. Overfishing and dirty seas alone would have been enough to send European oyster beds spiralling into decline, but one more factor sealed the deal: the remaining oysters were hit by mysterious and deadly diseases.

The demise of European oysters is not a one-off. Oyster habitats formed by dozens of different species used to reign in many places around the world but have been stripped away by a similar litany of ecological catastrophes. Globally, around 85 per cent of all previously known oyster beds, banks and reefs are gone. This decline is the average figure pulled from a large before-and-after dataset of 144 estuaries, mainly in North America, Australia and Europe, spanning the past 130 years. The actual losses could be even worse, because some of the early information was gathered when people had already begun to suspect that oysters were not doing so well. Surveyors were dispatched to see what was going on, often mapping the habitat by touch, feeling their way across the seabed to figure out where sharp shell reefs were growing. By the time they were surveyed, many areas had already started to degenerate from their former unspoilt state.

Despite that, no oyster species have gone extinct. They still live here and there, but only as sparse sprinklings across their former ranges. In Europe, almost all that remains of the giant oyster beds are piles of empty shells. That's what Andy Woolmer found in 2010 when he led a survey of 100 miles of coast along the south-western tip of Wales that points across St George's Channel to Ireland.

Andy's search area stretched between Swansea Bay and Milford Haven. Before setting out, he combed through archives of fishermen's records to see where oyster beds were historically known. He then went to have a look himself, to see what was still there. His team lowered video cameras into the water to search for signs of oyster life, and samples of seabed were scooped up and inspected. Among a smattering of living adult oysters, Andy found signs of new threats, including the recently arrived parasite *Bonamia*. These disease-causing microbes first arrived in Britain in the 1980s, and since then they've slowly been spreading around the coast, possibly carried by infected larvae. Most infected oysters appear normal, but under a microscope tissues from an adult oyster's heart or gills will show tell-tale signs of the spherical parasites. And four out of five infected oysters will die.

Another problem is a mollusc that doesn't naturally belong in British waters but hitch-hiked across the Atlantic when American oysters were imported in the 1880s in attempts to boost the failing native population. This is the Slipper Limpet, which lives clamped together with others of its kind in sex-changing piles and produces copious quantities of sticky goo (this is the limpets' pseudofaeces, the unwanted food they spit out before digestion). This smothers the seabed and prevents young oysters from settling. In other parts of the country – but not yet in Wales – a second invasive mollusc, the American Sting Winkle, has a greedy appetite for oysters. Having left behind their natural predators in their home waters, they tend to do rather well at the Native Oysters' expense.

As Andy found out, Welsh oyster beds were clearly in a sorry state, but he didn't think all was lost. Attending a workshop on the possibilities of oyster restoration and surrounded by scientists discussing the subtleties of oyster biology, Andy decided it was time to stop talking and start doing something: he wanted to see if he could put oyster beds back in the Welsh sea.

There were several reasons why he chose the Mumbles as a test site. It is one of the few areas the *Bonamia* parasite has not yet reached; it is also within reach of Andy's base in

Swansea harbour. Using high-resolution sonars borrowed from the Welsh Fishermen's Association, he scanned the seabed, and located several derelict oyster beds. These old piles of empty shells would form the spawning grounds for a new population. Working with experienced oyster fisherman Fenton Duke, he established the Mumbles Oyster Company.

'I wouldn't put my millions into oysters,' Andy tells me. Not that he has millions to invest, but he makes it clear that this venture is not about making money. His dream for the Mumbles is to create a fishery that is sustainable both economically and ecologically; that means a fishery that isn't reliant on external funds to keep going and one that will allow the oyster beds to flourish as an ecosystem for many years to come.

The team at the Mumbles Oyster Company want to bring back a mollusc and an industry that used to be a central part of the village's identity. Much has changed in the world since oysters were first fished off the Mumbles, and Andy is convinced that with twenty-first-century technology and thoughtful, progressive management they can make a go of it. He wants to show the rest of the world that the Native Oyster can come back from the brink.

It took a few years of paperwork for them to be granted access to a 35-hectare (90-acre) rectangle of seabed off the Mumbles pier. Once they had the go-ahead the first thing they did was to put some oysters down there.

During his surveys of the entire bay, Andy had found a handful of adult oysters and just two lonely juveniles, mini-oysters known as spat that were stuck onto other, empty shells. He needed to kick-start the oyster population.

With seed funding from the Welsh government and the EU, he bought 40,000 adult oysters from Loch Ryan in Scotland, one of the few remaining healthy, *Bonamia*-free populations of Native Oysters. Instead of being taken off to market, hundreds of bags of oysters were trucked down to the

Mumbles in batches, and during the winter 2013 to 2014 they were poured into the sea on top of the derelict oyster beds.

That winter turned out to be long, harsh and lashed by wild storms that left Andy anxious about his oysters down beneath the waves. So it was with a massive sigh of relief that, come the spring, when he returned to the oyster beds, he found a good number of the transplanted oysters had survived. The team had shown that Native Oysters can still live in Swansea Bay.

As I looked out from the end of Mumbles pier into the murky water I pondered the next and most important question. Are the transplanted oysters happy and healthy enough down there to start making more of themselves?

Adventures of an oyster

Oysters live complicated lives. Watching an oyster grow up, it almost seems as if it can't quite make up its mind about what it wants to be. It all begins, for the Native Oyster at least, when adult males cast clumps of sperm into the sea around them. All being well, some of the sperm will drift past a receptive female oyster who will draw them in through her siphon and use them to fertilise her eggs inside (in some other oyster species the females pump their eggs out into the water, and fertilisation is external). Problems in the lives of oysters can start right here. If there isn't a female nearby, those sperm will go to waste.

Sex for bivalves is never as intimate as the slimy clinch of snails, but even though they don't come into physical contact, mating oysters can't be too far apart. An oyster here and an oyster there is no use, which is why Andy went to the effort of bringing in thousands of breeding adults. Placing them in clusters on the seabed – around 10 per square metre – gives the breeding oysters the best chance of successful fertilisation. If this happens, the next step is for each female to brood embryonic oysters for up to 10 days among her gills and in her mantle cavity. During this time the youngsters are visible in a shucked oyster as a milky slick, referred to by some, revoltingly but accurately, as white sick. This gives at least one good reason why it is best to eat oysters only when there is an R in the month (a general rule that was first introduced by the Victorians in Britain); in the northern hemisphere these are the cooler times of year, September to April, when spawning isn't in full swing. Female oysters are quite harmless to eat during the breeding season but their soupy broods of larvae are perhaps not to everybody's taste. What's more it's a good idea, ecologically and economically, to leave oysters undisturbed during this time so they all have a chance to cast their offpsring into the next generation. As the days pass, the young oysters develop into grey and then black sick, by which point they are ready to leave their mother and fend for themselves.

Depending on her size, a single female Native Oyster will puff out as many as 1.5 million young into the water in one go. It's one of those facts that makes me think there really should be nothing in the world *but* oysters, but the ocean is a dangerous, difficult place, of course, and only a tiny fraction of those larvae will make it to adulthood.

Now autonomous, each baby oyster secretes a little shell that folds in two halves. It sprouts a brushy cluster of tiny hairs that waggle around and propel it through the water, and in this form it roams around for a few weeks until the time comes to add another piece of anatomy. The larva grows a foot that pokes out between the twinned shells like a tongue, and it sinks down to the seabed before tramping off to begin the most important hunt of its life.

Creeping along, the oyster tests out the substrate, looking for an ideal spot to settle down. If it doesn't like what it finds, the larva can launch itself back into the water column to catch a brief ride on a passing current and continue its search elsewhere.

What the adolescents are so desperately looking for is a certain smell or a taste that points the way to the ultimate prize: an empty shell to land on. Volatile compounds waft from living oysters as well as the vacated shells they leave behind, thanks to thin coatings of bacteria and other microbes. These invisible messages tell oysters to flock together. If there is no better choice they will make do and settle onto a stone or piece of wood, but they much prefer the scent of their own kind.

When the minute larva detects the right aromatic trail it crawls towards it and prepares itself for one final transformation. It comes to a halt, squeezes out a drop of chalky glue and cements its left shell in place on the outside of another oyster or onto an empty shell (the glue takes just a few minutes to set). From this point the larva is known as spat. Now, with no further need for mobility, the oyster reabsorbs its foot and grows huge gills which for the rest of its life – perhaps the next 20 years if it's lucky – will keep it flush with oxygen and particles of food.

Over the next 12 months the oyster spat grows into a mature male not much bigger than a thumbnail. Native Oysters start out life as males, then periodically undergo sex changes, gender-flipping several times during each spawning season, producing eggs, then sperm, then eggs again.

After three or four years the oyster will reach marketable size, seven centimetres (almost three inches) across, and if left alone in the sea for 15 years, it will continue to lay down layers of new shell until it reaches 11 centimetres or more. Long before then, its alluring smell will start drawing in other young oysters – including some of its own progeny, perhaps – so continuing the clustering of generations that builds up oyster beds and banks, spawn by spawn, year on year.

People have known about this part of the oysters' complex life cycle for a long time. Noticing that oysters of many species are inclined to huddle together and stick to shells, oystermen figured out a simple way of boosting their catches. Back in the golden era of oyster fishing in the US in the late

1800s, people collected up empty shells known as 'cultch' from shucking houses and canning factories, and threw them back into the waters they came from. As long as there were enough adult oysters left in the sea, these empty shells provided more places for their larvae to settle on and grow.

Laying down cultch was sometimes carried out in conjunction with the clearing away of oyster predators, including starfish. In New Haven, Connecticut in 1879, the 'starfish mop' was introduced. These frayed cotton ropes were dragged over the seabed to gather up starfish, snagging their sticky tube feet. The laden mops would then be brought up and dunked in vats of boiling water. Oystermen then dropped bushels of empty shells into the sea along with some adult oysters to help nurture the next generation. So, just like farmers on land, these seamen were farming the resources of the seabed.

The idea of putting shells back in the sea has been adopted more recently by conservationists, who are trying to undo decades of damage. Rebuilding ecosystems is a painstaking business that is not always successful, but for oysters it does seem to pay off. Restoration efforts, especially in the US, are beginning to show that fully functioning oyster habitats can be put back. Both sides of the country have their own species of habitat-forming oysters: Eastern Oysters on the Atlantic coast and Olympia Oysters in the Pacific. In dense aggregations they form solid reefs that can stand several metres up from the seabed, and in times gone by could be major navigational hazards. Back when oysters were super-abundant in the US they did more than tear open the occasional boat hull; they did a lot of good too.

When oyster reefs flourished they protected coastlines from storm floods and erosion; they nurtured young fish and shellfish that would eventually grow up, wander off and get caught and eaten by humans; and all those gaping bivalves played a vital role in keeping coastal waters clean and clear. There was a time, around 100 years ago, when every drop of water in many American estuaries, flowing from rivers and out to the sea, first passed across the gills of an oyster. Millions of oysters performed a crucial service, sifting and cleaning the waters, and all free of charge. Oysters remove suspended particles of mud and silt that can otherwise smother seagrasses and other sun-loving organisms. They also do a good job of slurping up excess nutrients from artificial fertilisers and sewage washing off the land and can curb outbreaks of harmful algal blooms. But in parallel to the European story of overfishing, pollution and disease, oyster reefs around American coasts faced a similar fate.

Today there is only one estuary in the US that is still known to have enough oysters to filter all of its water. That fact was uncovered by Philine zu Ermgassen from Cambridge University, who crunched a huge amount of data on oyster reefs past and present. Of 13 estuaries she studied, only in Apalachicola Bay on Florida's panhandle are there enough oysters left to filter all the bay's water before it pours into the Gulf of Mexico, and that incredible feat is largely thanks to the efforts of conservationists who have helped to put the oysters back.

Nearly every US coastal state has some form of oyster restoration programme under way. In many places, squadrons of willing local residents are volunteering their time because they want to see oysters growing once again on their watery doorsteps.

Various techniques for putting shells back in the sea are being tested. The shallow waters of Florida's Canaveral National Seashore, with the John F. Kennedy Space Center visible in the distance, are protected from dredging and fishing, but oyster reefs have still been suffering. The wakes from passing boats create bald patches where there used to be oysters. To help heal these gaps, 10,000 volunteers have hand-tied individual shells to plastic mesh mats. Among the helpers were cruise-ship crews, who spent their down-time at sea drilling holes in millions of shells ready to be fixed onto reefs.

Restored areas have since been overgrown by new oysters and are apparently indistinguishable from undamaged reefs.

In Louisiana and Alabama, living oyster reefs are being tried out as a means of protecting coastlines from hurricanes and storm surges. Recycled shells are tied up into mesh bags and pinned to the coasts, where they will gradually be overgrown by new oysters. Concrete reef balls are constructed like giant footballs with holes drilled in them and with oyster shells embedded to encourage more oysters to settle and reefs to grow.

Following America's lead, other countries are trying out oyster restoration. In the UK, plans are brewing to restore lost natives to several areas that had oyster fisheries in the past, including the Blackwater Estuary in Essex and the Solent on the south coast, but at the moment it's just Andy Woolmer and the Mumbles Oyster Company who are giving it a go, and a giant pile of empty shells is a key part of their efforts.

As well as transplanting all those mature Scottish oysters to Wales, the team have also added four tonnes of empty cockleshells to the derelict Mumbles oyster beds. There are already old dead shells down there, but Andy wants to know if adding more will help things along.

The empty shells came from the nearby Burry Inlet, where cockles are still gathered from the muds and sands at low tide using hand rakes and sieves, just as they have been for centuries. Owners of the cockle-processing plant let Andy help himself to the huge mounds of empty shells they produce. It still costs thousands of pounds to hire boats and crews to take the free cockleshells out to the Mumbles and Andy is working hard to find ways of making the whole venture economically viable.

One idea he has is to find uses for the invading Slipper Limpets. He is considering a trial system of retaining as many of these gummy intruders as possible when they come up as bycatch in his dredges, then freezing them to make sure they are quite dead, preserving them in salt and selling them to local anglers. Andy tells me they turn into little rubbery disks, which spring back into shape when soaked in seawater and apparently work well as bait for catching seabass and cod.

Restoring the Mumbles oyster fishery has become a three-pronged strategy: a new adult population has been introduced, shells have been added for newly hatched oysters to settle on, and unwanted molluscs are being removed. Now all the team can do is wait and see if their oysters will spawn successfully.

Oysters aren't the only molluscs that create ecosystems; many other species do their bit. Blue Mussels are a common sight in shallow coastal waters in the Atlantic and Pacific. They glue themselves to rocks and each other with sticky threads, and form wave-resistant beds. You might have seen them covering boulders and rocks at the beach.

Horse Mussels look like a larger version of Blue Mussels, although they don't taste as good, apparently. They colonise the seabed hundreds of metres beneath the waves, where they can live for 50 years. Solitary Horse Mussels are widespread and in just a few places they gather together and form thick carpets. Some of the most spectacular are in the Bay of Fundy in the Gulf of Maine where Horse Mussels pile up, forming banks up to three metres (10 feet) high, 20 metres (65 feet) wide and stretching for hundreds of metres. These habitats are highly vulnerable to the impact of dredging and may take decades to recover, if they ever do at all.

Perhaps some of the most surprising habitat-making molluscs are little clams called Flame Shells. They get their name from the bright ruffles of orange and red tentacles that stick permanently out from their shells (they can't close them all the way). Unlike adult oysters, which live their lives fixed in one place, Flame Shells can swim around, clapping their shells together and lifting off into open water when they get disturbed or feel threatened. Normally, though, when things are peaceful, Flame Shells get busy building nests.

In a similar way to mussels, each Flame Shell squeezes out a sticky net of silky threads that binds pebbles, gravel and bits of broken shell, forming a honeycomb structure that covers the seabed in a thick crust. The Flame Shells hunker down inside little tunnels, mostly keeping their gorgeous tentacles to themselves. When I first heard about Flame Shell reefs I imagined they unfurled an underwater red carpet that sets the seabed on fire. But in fact they are far more modest and secretive, and in a funny sort of way that makes me like them even more.

Dan Harries, from Heriot-Watt University's Scientific Dive Team in Edinburgh, knows Flame Shells well. He tells me they can be easy to miss. 'Occasionally they'll come to the entrance to their nests and you'll see them,' he said. 'But usually they're hidden away.' Instead, to get your eye in, you need to start looking out for suspicious bumps and lumps that shouldn't really be there on a flat plane of tide-swept gravel and sand. If you give the seabed a gentle prod, you'll notice it is soft and spongy.

Then there are all the animals that live among the Flame Shells. Thickets of bristle stars (leggy relatives of starfish) will congregate on a Flame Shell reef, thousands of them waving their arms in the water, while an occasional worm called a sea mouse will snuffle past with its luxuriant coat of iridescent spines. Sea sponges, soft corals and sea firs (relatives of jellyfish that look like miniature evergreen trees) are all devotees of Flame Shell reefs. In the midst of an ever-shifting substrate, they help themselves to the stable surfaces created by the reef where it would otherwise be impossible to get a grip. Clustered together, the clams and their nests transform the seabed from a featureless expanse into a bustling community.

A team of scuba-divers, including Dan, recently discovered the world's biggest Flame Shell reef. Dodging the vessels in a busy shipping lane, the divers sank down beneath the waves of the sea inlet, Loch Alsh, that runs between the Isle of Skye and the heaving backdrop of the Scottish Highlands. As the tides rise and fall each day they suck water in and out of the narrow channel that links the loch to the deep open water of the Outer Sounds of Raasay, making this an ideal spot for current-loving Flame Shells.

Dropping down and mapping out the Loch Alsh seabed, Dan and the dive team found Flame Shells *everywhere*. The reef they discovered covers roughly 75 hectares (almost 200 acres), an area equal to almost 3,000 tennis courts. It's home to an estimated 100 million Flame Shells.

'We were a bit curious as to why no one's noticed them before,' Dan admitted to me. A theory currently being tested is that these shelly reefs might naturally come and go. As the clams aren't cemented in place on the seabed, there's nothing stopping them upping sticks, moving on and building more nests elsewhere.

To create homes for other creatures doesn't necessarily require millions of molluscs, gathered together in great reefs and beds. Solitary seashells can also form important habitats. There are fishes and octopuses that lay their eggs inside empty seashells; on land, mason bees use snail shells as nests. The Belligerent Rockshell doesn't wait around for the other snail to die before turning its shell into a nest. Its victims are vermetid snails (known also as worm snails) that fix their tubular shells onto coral reefs like a random squeeze of toothpaste, with no mathematical elegance and with the open end peering up like an alien eye on a stalk. First the rockshell will suck the vermetid snail out of its shell, leaving behind a smear of eerie blue goo; it then turns around and lays its eggs inside the newly vacated tube. Charming.

Another group of animals that make use of second-hand shells are especially well known, perhaps because scientists and non-specialists alike find them endearing and fascinating in equal measure. These are crustaceans that seem to think they are molluscs, and have become experts at bringing empty seashells back to life.

Quietly watch over a tide pool and you might spy a seashell behaving strangely. Instead of sitting very still or perhaps gliding slowly and smoothly along, it scuttles in bursts, dashing forwards for a short way, then hunching down when it thinks danger is near. And if you pick up one of these curious shells there's a possibility that instead of an inquisitive soft tentacle peeping out you may be greeted by a sharp pinch.

Most crabs make their own shells. They construct a suit of body armour, which they shed and replace throughout their lives, each time finding somewhere safe to hide while their new outfit dries and hardens around them. Nevertheless, close to 1,000 living species of crab don't bother with that. They have permanently lost their shells, and have instead evolved ways to take advantage of empty seashells. These are the hermit crabs, and they've been borrowing shells for a very long time.

In 2002 an unusual fossil shell was found in Speeton, a village in Yorkshire, England not far from tall cliffs that overlook the North Sea. The shell was an ammonite, an extinct cephalopod that swam through far more ancient seas, in the Lower Cretaceous around 130 million years ago. After it died it sank down to the seabed where a crab scuttled past, picked it up and climbed inside. It was Dutch palaeontologist René Fraaije who spotted the perfectly preserved body of this hitch-hiker inside its ammonite home with its claws peeping out. This is the oldest known hermit crab, and the only one found inside an ammonite so far.

A naked hermit crab is a bizarre sight. It has a soft, extended abdomen that twists to a point, making it look like some sort of grotesque shrimp. A crab that lives inside coiled gastropod shells – as many species do – will push its bendy rear end into the empty shell and hold on tight, gripping the central pillar; it then retreats inside, plugging the shell opening with its claws, which have evolved to be just the right shape. These are trap doors that bite.

Other hermit crabs will grab on to a single, disarticulated bivalve shell – a clam or a cockle perhaps – and hold it over their head like an umbrella. Some specialise in long, narrow tusk shells. Their pincers are rounded to form a perfect plug for the entrance of their tubular homes. The one thing that hermit crabs never do is kill the occupant of a shell before moving in. They only adopt vacated shells and never consume their hosts: they wait for other animals to do that first.

The majority of hermit crab species live in the sea, and they have evolved a finely tuned sense of smell that draws them to the places where molluscs are being eaten. Particular peptides are produced when enzymes in predator-spit begin to digest mollusc meat; these waft through the water, and when a hermit crab picks up the scent, it marches off in search of a shell that will be abandoned any minute, just as soon as the predator has finished its dinner. Finding new shells is a critical part of being a hermit crab, and they devote a lot of time to this single pursuit. By not making their own shells hermits avoid the costs of construction, but it means that as they grow bigger they will keep on outgrowing their homes. Like Goldilocks, hermits are constantly on the lookout for the perfect shell: not too small, otherwise they won't fit inside, and not too big, otherwise the shell is too heavy and unwieldy to carry around.

Curiosity about how an animal evolves to rely entirely on the leftovers of another has led many scientists to watch hermit crabs very closely. These scientists are behavioural ecologists, a bunch who devote themselves to understanding why other animals do what they do. Those who specialise in hermit crabs tend to spend their time tinkering secretly with shells, numbering them, swapping them over, offering new ones and all the while watching how the crabs respond. From detailed behavioural studies one thing is becoming clear: hermit crabs do often live up to their name, and they can be quite antisocial.

For one thing, they have no qualms about stealing each other's shells. Suitable shells can be in such short supply that hermit crabs are permanently at risk of being evicted, and when two of them meet, a number of things can happen. The etiquette of a crab-to-crab encounter usually begins with a ritualised show of claws as the duo try to settle things without a fight. A larger crab will hold out its claws so its opponent knows exactly what it's dealing with (claw size is a good indication of overall body size and hence fighting ability and strength). This can sometimes end in surrender. The loser drops its shell and runs off naked; the victor can then take its time, inspecting the empty shell and perhaps trying it on for size before deciding if it wants to move house. If the situation is more evenly balanced, the slightly smaller crab might put up its dukes, thrusting its claws forwards repeatedly, probably in the hope that this will intimidate its aggressor and make it back off. But sometimes a scuffle is inevitable, and a hermit battle kicks off.

Crabs will wrestle each other, checking out whether their opponent's shell really is worth the effort. If it is, one crab will climb on to the other and repeatedly hammer the shell with its claws. Eventually, either the attacker runs out of energy and gives up or the defender has enough and relinquishes its shell.

Things are different for the dozen or so species of hermit crabs that live on land. Being high and dry, they accept that seashells are in especially short supply, and these hermits will sometimes have to make do with whatever they can find, perhaps a piece of wood with holes in or a discarded plastic bottle. In Madagascar, land hermit crabs have been seen waiting at the base of a crumbling cliff and picking up hollow fossil shells that occasionally drop out. On beaches, land hermits bustle to the tideline in the hope of finding a new home among the flotsam and jetsam, but there can be so many crabs around that most of the suitable shells will already be occupied. The severe housing shortage forces these crabs to socialise.

Whenever a land hermit crab is lucky enough to come across an empty shell (sometimes because a behavioural ecologist put it there) and if no one else is around, it will stop, take a closer look and probably try on the new shell for size. If it likes what it finds it will keep the new home and continue on its way. However, if the shell is too big the crab won't pass on by, but will sit quietly next to it, sometimes for as long as 24 hours. In that time other crabs will probably amble past and wonder what's going on. Then a spontaneous hermit party breaks out. Don't get too excited, though, because the main thing that happens when hermit crabs get together is they start forming queues.

A gaggle of hermit crabs clustered around a big empty shell will sort themselves out into a size-ordered line with the biggest at one end, leading to the smallest at the other. This orderly formation is called a vacancy chain, and people form them too, of jobs and houses. The crabs work out who goes where by clambering around and feeling up each other's shells. Sometimes, if there are lots of hermits in the area, several queues will form around a single, large vacant shell and then things get a bit more interesting: a tug-of-war ensues.

The biggest crabs will wrestle over the coveted empty shell while the little ones further down the line will shift queues like supermarket shoppers speculating on which checkout will move fastest. Eventually, one queue will win control of the empty shell and, in a flurry of claws, everybody in the successful line moves house. Each crab slips out of its old shell and into the newly abandoned shell of the crab one place ahead of it in the queue. They all get a new shell, one size bigger, and quickly scuttle off, once again going their separate ways. Behavioural ecologists have worked out that forming vacancy chains provides benefits for all the crabs involved; adding just one new shell can efficiently provide new homes, of just the right sizes, for a whole gang of hermits.

Behavioural ecologist Mark Laidre took on the enviable task of studying hermit crabs on the beaches of the Osa Peninsula on Costa Rica's verdant Pacific coast. In one experiment he coaxed the hermits out of their homes and gave them either new seashells or old ones previously worn by other crabs. From the outside, these two shell types seem to be similar in size, but second-hand shells have a larger entrance, and they're bigger on the inside because previous occupants have excavated them (they secrete chemicals that soften the calcium carbonate, then scrape away layers inside). When Laidre gave crabs new shells, they were usually too big to fit in and part of their bodies stuck out, leaving them vulnerable to attack by predators. In contrast, the hermits given previously occupied houses were mostly doing just fine tucked up inside their shells. As well as being more spacious, the remodelled shells are lighter and easier to carry around. Laidre put hermit crabs on little treadmills and measured how out much energy they use up carrying new and old shells. He found that crabs have a much easier time strutting around wearing second-hand shells.

The nub of the problem is that only the smallest, youngest crabs are able to move into new shells and begin the long task of digging out the interior, and they will only do this as a last resort, when they can't find a pre-used shell. On the beaches, there is a booming second-hand market in remodelled shells that become ecological heirlooms, passed on between many successive hermits.

In the sea, hermit crabs don't bother remodelling their shells. For one thing, seawater buoys up their shells, effectively making them lighter. Marine hermits also need their adopted homes to be as strong as possible to protect them from all the ocean predators, including plenty of other crabs, that have become specialists in cracking their way into tough molluscs. Excavating a shell and making it bigger, but weaker, just isn't worth the effort.

By keeping seashells in circulation and stopping them from getting buried and ground down into sand by the waves, hermit crabs are what ecologists refer to as ecosystem engineers. When beavers build dams and create ponds they are engineering ecosystems, as are woodpeckers drilling holes in trees, and European Bee-eaters digging nests in the ground and in steep cliffs; other bird species will move in after the bee-eaters have left. All of these engineers are creating, modifying and maintaining habitats that other species take advantage of. In the case of hermits, it's not just the crabs themselves that live inside the salvaged shells; they pick up other hitch-hikers along the way and become miniature, mobile ecosystems.

There are hundreds of species that creep and crawl inside hermit crab shells, or hang on to the outside and go for a ride. An up-to-date list of the things that live with hermit crabs goes on for almost 50 pages. They include worms that twist their way inside the shell and position their heads at the entrance, ready to steal morsels of food from the mouths of their hosts (they will also nibble hermit crab eggs). Sponges, sea squirts, barnacles, bryozoans, corals and shrimp all take up residence in and on hermit crab shells. There are even some gastropods that have flattened or concave shells that fit neatly inside a hermit's repurposed mollusc shell. Often two or three of these doubled-up shells will live together inside the same hermit cave.

All of these assorted hangers-on gain protection from predators and a hard surface to stick to, something that's generally difficult to come by on the boundless plains of soft, muddy seabed where many hermits roam. The hermits themselves also stand to gain from their cavalcade. Some will deliberately grab anemones and fix them to their shells, making a stinging line of defence; they will even bring their favourite anemones along with them when they move to a new house.

One unusual variety of sea anemone helps out by building an extension to the hermit's shell as it grows. *Stylobates* was originally found in 1895 by American naturalist William Healey Dall, who identified it as a peculiar type of deep-sea snail. Twenty-five years later he changed his mind and realised the golden spiral was in fact an anemone that grows around a snail shell; it makes a gleaming, papery model of the shell, rather like covering a balloon in layers of newspaper to make a papier-mâché bowl. When the anemone reaches the open mouth of the shell it carries on growing in the same spiral shape, as the snail did when it was alive. The anemone gains a solid, secure place to stick to, and the hermit crab will never grow too big for its shell. It seems to be an ideal partnership.

It's several months before I hear from Andy Woolmer again. The transplanted oysters are settling in for their second winter off the Mumbles, and Andy has been out to check on them. He emails, telling me about the video cameras he dropped down, but he couldn't see much through the silt and minerals washing in from the nearby Tawe and Neath Rivers that turn the water into a flocculated snowstorm. He did a few tows with the dredge and brought up a good collection of healthy Scottish oysters that still seem to be getting on well in their new lodgings in Swansea Bay. Some were growing bigger with a white frilly edge to their shells, a line of new growth.

I click to open a photograph Andy has sent me. On my screen I see a hand holding a large adult Native Oyster. Stuck to it is another, much smaller oyster: an oyster spat.

Andy tells me he can't be quite sure that it was born in Wales, and hadn't been brought down from Scotland already clamped to the adult's shell. Either way, it's definitely a sign of good things to come for the return of these oysters, which have been missing from the Welsh coast for such a long time.

Clockwise from top: A sacoglossan sea slug beside its spiralling (Archimedean) egg ribbon; Bobtail Squid with its flamboyant, communicative mantle on display; Lined Chiton, its shell made of eight overlapping plates; a gaping Giant Clam.

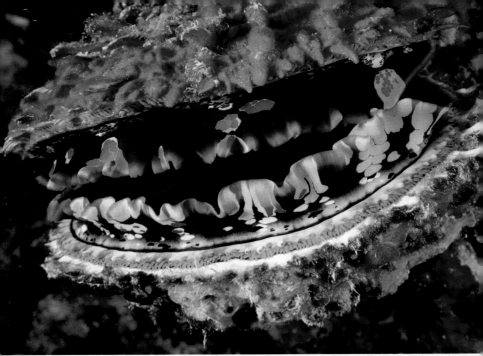

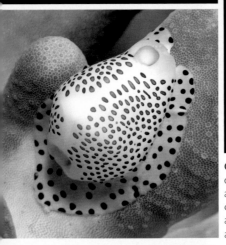

Clockwise from top: A *Spondylus* thorny oyster; an Aztec double-headed serpent made of turquoise, red *Spondylus* and white conch shell, probably part of a sixteenth-century ceremonial costume; a Little Egg Cowrie; mask made from a large *Spondylus* shell in Manabi, Ecuador between 800 and 400 BC.

Clockwise from top left: Andy Woolmer pouring oysters into the sea off the Mumbles in Swansea Bay, Wales; a handful of Native Oysters ready to go back to the seabed; a Striped Hermit Crab. It has made a *Calliostoma* top shell its home, with a fringe of stinging hydroids clinging to the outside.

Clockwise from top left: Fatou Janha cheers on wrestlers at the Gambian Oyster Festival; an Oyster festival costume; a midden of Gambian oyster shells; a member of the TRY Oyster Women's Association shucks oysters; celebrations at the festival.

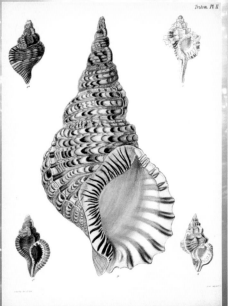

Clockwise from top left: *Triton* shells from the *Conchologia Iconia*, drawn by Lovell Reeve and based on shells at the Cuming Museum, 1843; Blue-ray Limpets, clustered on a kelp frond; the teeth of a Common Limpet seen under an electron microscope. These teeth are made of the strongest biological material known – all the better for scraping the limpet's algal food from rocks; a chambered nautilus, swimming in the sea off the island of Palau in Micronesia.

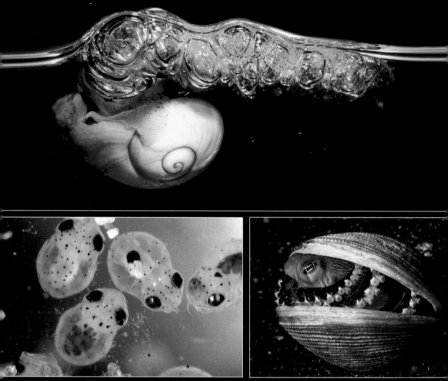

Clockwise from top: A *Janthina* snail floats at the surface on a raft of bubbles, camouflaged against the open ocean by its blue shell and foot; a Veined Octopus peers from the bivalve shell that it uses as a hideaway; a female argonaut peeps from her shell, which she uses as a portable chamber to brood her young and control her buoyancy; a clutch of baby argonauts, each around 1mm long.

Clockwise from top left: Raw byssus from a single Noble Pen Shell; sea-silk embroidery by Assuntina and Giuseppina Pes; Ignazio Marrocu demonstrates a tool that was once used to harvest pen shells, at the Museo Etnografico in Sant'Antioco; a Noble Pen Shell, standing high above the Mediterranean sea bed; sea-silk weaver Efisia Murroni.

Clockwise from top: Newly discovered micromolluscs from islands off Papua New Guinea, found by Philippe Bouchet and his team; KOSMOS mesocosms in Gran Canaria, used to study the effects of increasing acidity on open ocean ecosystems; a sea butterfly with its tiny wings and left-coiling shell; a sea angel – not as angelic as it appears. This shell-less swimming gastropod is a deadly enemy of the sea butterfly.

CHAPTER SIX

Spinning Shell Stories

Many stories have been told about a strange and fabled cloth called sea-silk. Some say that when Jason and his troop of mariners set sail aboard the *Argo* they may have been hunting for a golden fleece made of sea-silk. There are stories of Roman emperors bearing robes trimmed in shimmering sea-silk, with dancing girls clad in see-through dresses of this same fabric that apparently left little to the imagination. A pair of gloves made of sea-silk has been said to be so dainty, they fit inside half of a walnut shell. Ancient Egyptian kings had boats powered by sea-silk sails, and Egyptian mummies were thought to be wrapped in sea-silk cloaks. Sea-silk was commonly associated with the 'cloth of gold' mentioned many times in the Bible. When Henry VIII met the King of France in 1520 at the 'Field of the Cloth of Gold', the story runs that the field was decked out with sea-silk flags and bunting, with Henry's men dressed to match in fine golden tunics. As for the source of this fine fabric, a peculiar set of stories emanated from Chinese traders in the second and third centuries. Water sheep, they said, lived beneath the waves in the Roman Empire and occasionally clambered onto the shore, where they scratched themselves against rocks and left behind clumps of wool; people gathered these tufts and wove them into fine cloth. Similar stories emerged among Arab traders in the tenth century, who told of a beast called *abu qalamun* that would emerge from the sea at certain times of the year and shed its golden hair along the shoreline. This hair was made into a cloth so rare and valuable that its export was forbidden. Later, a twelfth-century Moorish writer declared the source of these fibres to be a creature that resembled a small sheep with webbed feet like a duck.

All this may be starting to sound rather far-fetched, and it is likely that the water-sheep stories were really just a joke that got a bit out of hand. From Roman times onwards, other writers based closer to the Mediterranean mention another possible source of sea-silk. They wrote about fine silken threads that came from giant seashells with gleaming beards. It's here that the stories of sea-silk begin to edge closer to the truth.

Since antiquity, a large species of Mediterranean bivalve has gone by the name of the *pinna*. Today they are called Noble Pen Shells, *Pinna nobilis*. They look like huge mussels standing alone on the seabed, at least as wide as a man's outstretched hand, up to a metre (three feet) tall, and often covered in a fleecy cloak of seaweeds. They can live for 20 years or more, and while there are several other *Pinna* species, none are as large as the Noble Pen Shell; these are the biggest seashells in the Mediterranean.

A net of silky threads with sticky ends sprouts from the shell of this towering mollusc, to stop it from tumbling over in brisk underwater currents; the threads root the Noble Pen Shell to the seabed. Other bivalve species produce similar strands; if you have ever cooked mussels you may have had to clean them first, pulling off their mossy beards.

These fibrous anchors are formed in a process similar to the production of injection-moulded plastics. An internal gland secretes liquid collagen proteins that trickle along a groove in the mollusc's foot. The proteins take a few seconds to set hard into a narrow strand while the mollusc presses its foot against the seabed; an adhesive pad at the end of each thread then sticks to seagrass roots, sand grains or other fragments in the seabed. Once the new thread is ready the mollusc lets go, and it will continue making more until it has a beard of 1,000 or so hairs poking out from its shell, and fastened to a central stem lodged deep inside its body. Roughly the width of a fine human hair, the threads can be up to 20 centimetres (eight inches) long. They are known as 'byssus', a word often used for sea-silk. Are these delicate filaments the source of the ancient golden fabric? The answer to that is yes. And no.

It was an American biologist and science historian, Daniel McKinley, who in the 1990s decided to try to find out exactly how pen shells came to be pulled up from the depths and thrust into so many myths and fables. He picked up many strands of sea-silk stories and followed them back in time to see where they began. Through hundreds of manuscripts, books and museum specimens, he hunted for evidence to separate the truth from accumulated layers of mythology. What is sea-silk? When people wrote about pinna shells and byssus, what did they mean? Have these fine fabrics really been around for thousands of years? McKinley gathered together his findings and in 1998 published a monograph called *Pinna and her silken beard: a foray into historical misappropriations*, which already gives you a good idea of what he had to say.

A major snag in the sea-silk stories that McKinley encountered is the changing meaning and spurious translation of
words. The modern-day meaning of the word 'byssus' is
clear-cut. The fibres many bivalves use to fix themselves in
place on the seabed are called byssus, and they are made by
the mollusc's byssus (or byssal) gland. It follows that a fabric
woven from those filaments should also quite reasonably be
called byssus. The problem is that the term hasn't always
referred specifically to fibres made by molluscs. Tracing the
word 'byssus' back in time, McKinley saw that the solid
definition begins to get hazy until all certainty evaporates,
and it becomes impossible to know what writers were
actually writing about.

Similar words in several ancient languages including Latin,
Greek, Hebrew and Phoenician were used as general terms
for a range of fine cloths that could have been made from
linen or cotton or sometimes silk; the particular material is
not always specified. In the Old Testament, for instance, the
Hebrew words *būṣ* and *šeš* have been variously translated at
different times into the Latin word *byssus* as well as 'fine
linen' and 'silk' in English and *bisso* in Italian.

An important waypoint in the story of byssus is Aristotle.
He was supposedly the first person to connect the word
'byssus' with Noble Pen Shells and their luxuriant beards.
However, when we delve into the details of what he actually
wrote, and how his words have been translated, a different
story emerges.

In his book *The History of Animals*, written in 350 BC,
Aristotle mentions *pinna*, and numerous translations have
been made from the original Greek. In 1910, for example,
zoologist and shell-shape ponderer D'Arcy Wentworth
Thompson translated some of Aristotle's text as 'the pinna
grows straight up from its tuft of anchoring fibres in sandy
and slimy places'. Much earlier, a thirteenth-century Latin
translation described the shells as growing 'upright out of
the depth in sandy places'. This phrase comes from Aristotle's

original Greek word, βυσσου, interpreted in this instance as βυσσός, meaning 'depth' (from which the words 'abyssal' and 'bathysphere' stem). This is probably what Aristotle was originally getting at (where Thompson got his 'slimy places' from isn't clear). The real problems arose in the second half of the fifteenth century, when Theodorus Gaza, a Greek translator living in Italy, undertook a major re-write of Aristotle's book. One thing he changed was the translation of that key Greek word. Instead of 'depth', Gaza read βυσσου as 'byssus' or 'fine linen'. The difference comes down to the subtlest of slip-ups, shifting an accent from the last syllable (βυσσός) onto the first (βύσσος), which transforms its meaning from one word to the other (accents were a later addition to Greek that weren't used in Aristotle's time). And so, as easily as that, the pen shells were now growing upwards from their fine byssus, much like a tree growing up from its roots.

Gaza's translation of *The History of Animals* was published in 1476 in Venice, and it was immensely popular, far outselling all the previous versions. By making this misleading connection between pen shells and byssus, though, he sparked a game of Chinese whispers that has gone on ever since. Stories were reshaped and new ideas became fixed until most writers and historians uncritically came to assume that any mention of byssus, no matter how far back in the past, could have referred to sea-silk woven from the Noble Pen Shell's fibres.

The true story, now well hidden and seldom told, is that up until the fifteenth century there was no reason to link byssus and pen shells. All the various ancient mentions of byssus – in the Bible, on the Rosetta Stone, on ancient papyrus scrolls and elsewhere – most probably referred to linens, or mulberry silk made by moths.

Given all this, Daniel McKinley remained sceptical about many of the ancient stories of sea-silk. He was sure that the idea of Jason and the Argonauts chasing after a fleece made of sea-silk, however tempting, was just one of many embellishments added to the myth throughout centuries of storytelling. Analyses have shown that Egyptian mummies are wrapped not in sea-silk but in linen. And in McKinley's view, the links of sea-silk to the biblical cloth of gold were equally shaky; Henry VIII and his men almost certainly never dressed head-to-toe in sea-silk.

Nevertheless, sea-silk *has* been around for a long time, although not as widely or with as much significance as many still claim. In reality, sea-silk has always been incredibly rare.

From myths to reality

The earliest authentic written mention of sea-silk, one not based on hearsay or mistranslation, comes from the turn of the third century AD. 'Nor was it enough to comb and sew the materials for a tunic. It was necessary also to fish for one's dress.' This quote is attributed to a man known as Tertullian, from Carthage in the African provinces of the Roman Empire. He goes on to describe how fleeces are obtained from 'shells of extraordinary size' that have tufts of mossy hair. He was clearly talking about pen shells and their byssus beards.

Sea-silk is one of the commodities listed by the Roman Emperor Diocletian in a price-fixing scheme that he rolled out across the empire in 301 AD, to try to stop merchants from fleecing their customers. Sea-silk crops up again in Constantinople in the middle of the sixth century when Emperor Justinian handed out gifts to visiting dignitaries including a 'cloak made of wool, not such as produced by sheep, but gathered from the sea'.

As for actual remains of ancient sea-silk, these are even more fragmentary and hard to find than written words. While we could blame clothes moths for eating the evidence, other natural fibres are just as vulnerable to getting munched and yet they show up much more frequently in the archaeological record. The oldest known piece of sea-silk dates from more than 1,700 years ago in the fourth century. It was found in Budapest, in the remains of what was formerly a Roman legionary town called Aquincum on the northern fringes of the empire. In 1912, a grave was found there containing a female mummy wrapped in linen. Between her legs was a fragment of fabric identified at the time as sea-silk. It was described as being coarse and brittle and as if it was made from human hair. Under a microscope, the cut ends of the fibres were seen to be egg-shaped, a unique feature of sea-silk. It remains unknown where this scrap of fabric was made; the piece was lost amid the chaos of the Second World War.

To find the next oldest piece of sea-silk, and the oldest surviving and scientifically verified example, we have to jump forwards in time 1,000 years to the fourteenth century. A knitted hat was excavated in 1978 from a damp basement just outside Paris. It has a few holes in it now, but you can clearly make out that it was a close-fitting beanie hat. The idea that sea-silk was flimsy and delicate doesn't quite ring true with this piece of clothing; warm and woolly are the words that spring to mind.

In his book, Daniel McKinley hunted for proof that sea-silk fibres had ever been woven or knitted into chiffony fabrics, and he drew a blank. Stories of sea-silk gloves kept in a nutshell may be yet another mix-up, this time with an early nineteenth-century trend for so-called Limerick gloves. Made in Ireland and Scotland from fine leathers, they were indeed sold stuffed into walnut shells.

The idea that sea-silk can be quite cosy fits with a rare literary appearance of this elusive fibre. In *Twenty Thousand Leagues Under the Sea*, Jules Verne dressed the renegade explorer Captain Nemo and the crew of his submarine the *Nautilus* in uniforms made of byssus. At the start of the book Nemo kidnaps the scientist, Professor Aronnax, whose expedition attacked the *Nautilus* thinking it was a dangerous sea monster. Nemo and his captive crew then venture around the globe exploring the underwater realm and, at one point, they cruise close to a submerged volcano; conditions on board become so hot that Aronnax feels obliged to take off his byssus coat. In the original French version of the book, Verne goes to some lengths to describe what he means by byssus, explaining that his submariners harvested fibres from pen shells to make their clothes. These details are skipped over by translators in many English editions, leaving readers to ponder the contents of Nemo's wardrobe.

I began to suspect that seductive dancers of the Roman emperors would have been thoroughly disappointed by what sea-silk had to offer when I saw a piece of it for the first time. I was visiting the mollusc section at London's Natural History Museum; curator Jon Ablett met me in the museum's great entrance hall, beneath the iconic *Diplodocus* skeleton, and led me through a door and down a set of narrow stairs to the back rooms that house the bulk of their enormous collections. Molluscs alone take up several huge rooms and long corridors lined with wooden cabinets; Jon opened a drawer in one. Pulling out a small box, he showed me a golden-brown glove. It's one of four sea-silk gloves that belonged to Hans Sloane, the man whose seventeenth-century collection formed the foundation of the British Museum, and in time its natural history division. I wasn't allowed to try it on but the glove looked to me to be rather thick and itchy, not gauzy and delicate; you would definitely be hard pressed to find a walnut big enough to keep it in.

The glove is one of around 60 items listed in a catalogue of all known sea-silk objects. Project Sea-silk is based at the Natural History Museum in Basel, Switzerland, where its founder and sea-silk scholar, Felicitas Maeder, is gathering records and information about sea-silk, all of them available to see on her website. She has scoured museum collections around the world for items made of sea-silk from before the 1950s. Knitted gloves and gauntlets are the most common items Felicitas has archived, along with a few hats, scarves and ties. Tufts of golden sea-silk have also been made into unspun fur. The Field Museum of Natural History in Chicago has an Italian muff in its collection and the Musée Océanographique in Monaco has several furry sea-silk objects including a lady's purse that looks rather like a Scotsman's sporran.

Most of the objects in the Project Sea-silk archive date from the eighteenth and nineteenth centuries (the fourteenth-century Parisian hat is one of a kind), and many of them were made in Italy. It was around this time in the southern Mediterranean that the stories of Noble Pen Shells and sea-silk began to untangle, and a clearer picture of this legendary fabric emerged.

'They tell me they are very scarce, and for that reason I wish you to have them.' These were the words of Horatio Nelson in 1804, a year before he died at the battle of Trafalgar, written to his lover Emma Hamilton. He was referring to a pair of gloves made 'only in Sardinia from the beards of mussels'. By that time, fine items of sea-silk like Emma's gloves were becoming more familiar.

The origins of sea-silk remain stubbornly mysterious, and no one knows for sure who first thought to pluck hairs from giant seashells and turn them into threads and fabric. Certainly by the Renaissance, Noble Pen Shells and samples of sea-silk began appearing in cabinets of curiosities.

Scholars and noblemen across Europe developed the habit of curating private collections of assorted objects and oddities. Both natural and man-made curiosities were displayed side by side in specially made pieces of furniture, or spilled over into entire rooms: stuffed animals and skeletons, feathers, butterflies, seashells, corals, bits of old pottery, shrunken human heads, coins, even unicorn horns and mermaids, which were often covertly cobbled together from an assortment of real animals.

The idea behind these collections was to assemble a physical encyclopaedia that helped make sense of how the world worked by drawing connections between apparently quite different objects. They arose before science and art were pulled firmly apart and assigned their own distinct disciplines. Onlookers would have no doubt marvelled at sea-silk, and puzzled over where it came from.

By the nineteenth century, sea-silk was being put on display at international exhibitions as an example of fine craftsmanship. Sea-silk appeared at the Louvre in Paris in 1801, and in 1876 was brought to America and displayed for the first time, at the Centennial Exposition in Philadelphia that celebrated 100 years since the signing of the Declaration of Independence.

Accounts of how these sea-silk items were made, and where, come from a coterie of early travel writers, mostly young gentlemen who went on Grand Tours of Italy. According to these sightseers, fishermen along Italy's Mediterranean coast used long metal tongs to probe the depths for pen shells; divers also swam down, tied ropes around them and yanked the shells back up to the surface. It was mostly women, especially in nunneries and orphanages, who took on the task of washing, combing, spinning and finally knitting or weaving the fibres together. As one writer in 1771 noted, 'The preparation is both laborious and ingenious.'

The centre of the sea-silk industry is pinpointed in many reports in Taranto, a city on the southern tip of Italy, inside the heel of its boot. Some confusion remains over a fine fabric called tarantine also made in the city, which some say could have been sea-silk, but was probably in fact made from fine sheep's wool (regular, terrestrial sheep that is, not water-sheep). Other mentions of sea-silk come from Naples, Sicily and Corsica, as well as Spain and mainland France, but the only other place where its production has been firmly identified is Sardinia.

By all accounts the sea-silk industries in Taranto and Sardinia could never have been very big. Nelson hit the nail on the head when he described Emma's gloves as being incredibly rare. For one thing, the supply of byssus threads was, all things considered, quite tiny. To knit a single pair of gloves probably required 150 shells, and unlike a field of cotton or a herd of sheep that can be harvested and shorn many times, pen shells would have produced only a one-off haul of material; they were brought up from the depths and killed for their beards. People sometimes ate the meat, too. Greek and Roman writers had mixed feelings about how good *pinna* was to eat, saying it was difficult to digest and diuretic, although the meat from smaller shells was apparently tasty when marinated in wine and vinegar. In southern Italy, *pinna* was cheap food until fairly recently, with various recipes including frying them in breadcrumbs, boiling them into broth, cooking them in lemon juice and serving them with baked prunes.

Another hint that sea-silk production never exactly flourished comes from reports of people who endeavoured in vain to stimulate the industry. In the 1780s, archbishop Giuseppe Capecelatro hoped to create jobs for impoverished sea-silk weavers in Taranto. He tried to kindle demand for the fabric by handing out sea-silk gifts to visiting dignitaries. In the mid-nineteenth century, Sardinian doctor Giuseppe Basso-Arnoux remembered his childhood Sundays, when his family dressed in fine sea-silk accessories, scarves and gloves. Later in life he decided to try to bring back these traditions. Visiting London, he attempted to establish a trading interest in sea-silk, but as with Capecelatro and anyone else who tried, his efforts never amounted to much.

More recent attempts have been made to rejuvenate sea-silk manufacture. In Taranto in the 1920s, Rita del Bene tried and failed to establish a government department of sea-silk, so instead set up her own private school to teach the craft, which continued with some success until the outbreak of the

Second World War. An interest in sea-silk in Taranto never revived after peace returned to the region. However, the processing of sea-silk has not disappeared altogether.

To the west of Taranto, 120 miles across the Tyrrhenian Sea on a tiny island off the coast of Sardinia, the craft clings on. It was there that I tracked down the trail of the sea-silk, finding a place where a few strands of this mythical thread are still made, and plenty of stories are still told.

The directions to Sant'Antioco read like something from a fairy story: drive down the road lined with prickly pear trees, go past the flock of pink flamingos and carry on over the bridge leading to a little island. There you will find the only people in the world who still pluck tufts of hair from giant seashells and weave them into fine golden cloth.

Bumbling along in the Fiat Cinquecento I hired at the airport, I slow down to catch a glimpse of the orange and yellow houses clustered on the hillside, overlooking an outrageous blue sea that I am told is the hiding place of Noble Pen Shells. I had come to meet the women who hold on to the secrets of sea-silk, and uncover what truths I could about this most mythical of fabrics.

At the top of the hill, above narrow cobbled streets, there is a high wall surrounding an open courtyard and a small, stone building where grapes were once processed and made into wine. Now the space is home to a collection of tools and machinery that have been used in Sant'Antioco over the last few centuries. This is the Museo Etnografico, run by a local cooperative called Archeotur whose members are committed to making sure past lives and traditions are not lost in the melee of modern life and that people don't forget how things used to be. Preserved in this modest space is an archive of local trades, of bread-making, cheese-making, shoe-making, barrel-making and the dyeing and weaving of local fibres including sea-silk.

Waiting to welcome me in is Archeotur's director, Ignazio Marrocu, a smiling man with a silver moustache and bright pink shirt. He immediately whisks me over to a cluster of Noble Pen Shells, standing tall and empty in a glass tank of sand. He pulls one out and hands it to me. The shell is at least 50 centimetres (20 inches) long, and surprisingly heavy. At the open end, the part that would have stuck up above the seabed, the pen shell is covered in the twisting white casements of tube worms and dried strands of seaweed; the lower section tapers to a point, and is scaly like reptilian skin.

Next, Ignazio brings out a knotty tangle of threads embedded with tiny seashells and blades of seagrass, like the ginger beard of an old man of the sea, flecked with his dinner. This is the byssus from a pen shell in its raw, untreated state. He then places in my hand a tuft of soft golden fibres that gleam in the sunshine. This is clean and carded byssus, ready to be spun. This is sea-silk.

The museum has a large display board covered in photographs of sea-silk weavers of the past. One black and white picture depicts four young women sitting in a row wearing headscarves, long dresses and aprons; one has a basket on her knee, filled with a tangle of byssus; the other three have wooden spindles and are in the process of twisting the fibres into threads.

Another photograph, this one in colour, shows an old lady wearing big round glasses, a white headscarf and a blue dress. Like the girls in the older picture she is busy spinning sea-silk. This, Ignazio tells me, is Efisia Murroni, who died in 2013 shortly after her hundredth birthday. She had learnt how to weave sea-silk from Italo Diana who ran a studio in Sant'Antioco, weaving traditional Sardinian designs and textiles until his death in 1959.

Surrounding the photograph of Efisia are pictures of Italo's work. There is a woven hat and jacket for a toddler, a wide knitted scarf with golden tassels and an embroidered tapestry, as tall as the women holding it up. The intricate design has a pair of horses (or possibly unicorns), and a pair of birds that look like fancy turkeys. Around them is a border of other animals, and a row of people holding hands. In the centre is a rather confused patch of stitches, one that tells a story of how the piece was made.

Italo wove and embroidered this piece in the 1930s for the occasion of Benito Mussolini's visit to the nearby town of Carbonia. It was a new town, built around a coal mine (*carbone* meaning 'coal' in Italian), and the streets were laid out in the shape of the egomaniac Mussolini's face. The central piece of the tapestry had originally been the words 'Il Duce', but this embroidered tribute to fascism was later covered over with new stitches.

Italo's skills have been passed on via Efisia not to her daughter, who didn't want to learn, but to two other women from Sant'Antioco. Several years ago, Assuntina and Giuseppina Pes became interested in the town's traditions of weaving sea-silk, and Efisia agreed to teach them.

The Pes sisters arrive at the museum, after dropping off their children at school, and greet me with smiles and cheek kisses. They are keen to show me their sea-silk skills, so we jump into an aged BMW driven by Giustino, one of Archeotur's enthusiastic volunteers, who knows English better than I do Italian. We zoom off to the outskirts of Sant'Antioco and pull up to a little house guarded by a friendly, yowling cat.

Assuntina opens the door and ushers us into her home, where bright sunshine pours into a room crammed with two large weaving looms draped in skeins of brightly coloured wool. The walls are decorated with weavings and embroideries of traditional Sardinian motifs. She leads us downstairs into a smaller, darker room and brings out a large plastic Ferrero Rocher chocolate box packed with plastic bags; she then lays a small collection of byssus out on the table. Together, Assuntina and Giuseppina set about showing me the stages involved in making sea-silk.

The first piece is byssus after it's been soaked for hours in seawater, then freshwater (at this point it hasn't changed too much), and it is beginning to be transformed, with the sandy, shelly debris picked out. Assuntina opens a red cardboard box with a puff of fibres inside that resemble auburn human hair. She grabs a handful and combs them over and over, teasing them with a fearsomely spiky comb. It reminds me of the painful brushing of my tangly, curly hair each morning before school.

Now, she takes out a wooden spindle, the kind used to spin cotton, wool and linen threads. It looks like a mushroom with a long, narrowing stalk and a small hook on top. She attaches a clump of combed byssus fibres to the hook and sets it spinning. I watch as the spindle spins and twists the byssus into a thread that wraps around the stick. Assuntina deftly feeds the growing thread with more fibres, making it look easy, but I know it isn't.

In a few minutes she spins a metre or more of thread. It is fairly thick and woolly, but soft to the touch. She tells me that the threads can be soaked in lemon juice to give them a brighter colour. One of their intricate embroideries features a pair of birds gazing at each other, beak to beak. They are sewn onto white linen with byssus of two different shades, one a deep bronze, the other pale gold.

As well as using sea-silk as an embroidery thread, it can be woven into fabric. A tiny tabletop loom comes out and Giuseppina shows me a narrow sea-silk tie in progress. I imagine their grandfathers dressed up in ties like this for church on Sundays. With her fingers nimbly darting this way and that, Giuseppina runs the golden-brown weft thread across the warp and pats them into place, making one more row of fluffy cloth.

No one will wear this tie, and it may never be finished, because new byssus fibres are very hard to come by these days. At the museum, Ignazio had demonstrated for me a metal tool with a long wooden handle that was used to wrench pen shells up from the shallow seabed, a few feet deep, but that is no longer allowed. Since 1992 there has been a blanket ban on harvesting Noble Pen Shells.

Along with seahorses, otters, seals and more than 200 other European species, Noble Pen Shells are protected throughout their ranges under EU law. Scientific advisors declared that pen shells are threatened by pollution and the destruction of seagrass beds where many of them live. Pen shells are easily crushed and torn away by boat anchors and fishing gear; also, divers were collecting them not for their byssus but to make the shells into gaudy home decorations, lampshades and the like. Now it is a criminal offence to deliberately harm or kill a Noble Pen Shell.

With the pen shells protected, Assuntina and Giuseppina see no way to obtain sea-silk, but it's something they seem calmly resigned to. It's clear they would both like to preserve the skills passed on to them from Efisia and Italo, but all they have is a dwindling collection of old byssus fibres handed on to them. Occasionally a local fisherman will find a dead pen shell and give it to the women to use. Even so, their byssus stock is small, and sea-silk is becoming rarer and more precious than ever.

The Pes sisters are not alone in continuing the traditions of sea-silk. Patricia, another member of Archeotur, has come with us to watch them at work and in a lull in the conversation she smiles and softly says something in Italian. Giustino translates for me. 'She says her grandmother weaves sea-silk too.'

We all say goodbye and Giustino drops me off in town, where I pay a visit to another of Sant'Antioco's sea-silk weavers, one who has something the Pes sisters and Patricia's grandmother don't have: a ready supply of new byssus.

I step into the cool, dim interior of the Museo del Bisso –
the Byssus Museum – and instantly feel as if I have walked
into the fairy tale that my journey to the island had promised.
This vaulted stone room was once the town's grain store
and is now a shrine of sorts to sea-silk as well as to the
woman who calls herself the last surviving maestro of sea-
silk, Chiara Vigo.

The walls are lined with glass cabinets containing a myriad
of puzzling objects; a bronze sculpture of a pen shell (far
bigger than the real thing) stands on the floor; there are
giant portraits of Chiara, and a huge undersea diorama of
fish and shells and mermaids. A small congregation sits in
hushed silence on chairs lined up in front of Chiara's table,
where she is busy at work.

A great deal of mysticism surrounds spinning and weaving,
especially female weavers. Sleeping Beauty fell into a deep
sleep after pricking her finger on a spinning wheel. Alfred
Tennyson's Lady of Shalott, based on Arthurian legends and
depicted in many Pre-Raphaelite paintings, was under a
curse that meant she couldn't gaze directly at the real world
but could only weave the 'half shadows' she saw reflected in
a mirror. In Roman and Greek mythology, a trio of goddesses
would spin, measure and cut the threads of life. Legends
around the world bestow great power, wisdom and magic
on women who weave. I find a seat in the Museo del Bisso,
next to Rebecca who has come to help translate for me, and
I can't help thinking this place endeavours to channel those
same time-worn enchantments.

Illuminated by a bright table lamp, Chiara is carrying out
the same meticulous steps of combing and spinning the
byssus threads that I saw at the Pes sisters' house, though
Chiara adds her own particular twists to the proceedings.
While Chiara works on her strand of sea-silk she tells a
stream of stories. She tells her onlookers about the ancient
origins of sea-silk in the Middle East, 10,000 years ago; she
tells of sea-silk in the Bible, and the source of King

Solomon's shining robes; she tells of her personal oath sworn to the sea.

Chiara plays a game I imagine she repeats many times a day, asking me to hold out my hand and close my eyes. I feel nothing and open my eyes to see a weightless cloud of sea-silk threads sitting on my palm.

Now picking up a wooden spindle, she begins to twist the fibres together, and while she does she sings a song. I don't ask Rebecca to interpret the words of the Italian sea shanty but I listen to the tune, and Chiara smiles a twinkling smile at her transfixed crowd as the byssus spins round and round. Someone in the audience joins in with a few lines of the song.

When the pile of byssus fibres have all been twisted into one long thread, Chiara unwinds the spindle and brings out a white plastic cup half full of a pale yellow liquid. She explains this is a special mixture – a secret recipe – of lemon juice plus extracts from a dozen different seaweeds and the juice of another large Sardinian fruit. Chiara dunks the byssus thread into the liquid then draws it out, squeezing and dabbing it gently with a tissue. Then for the first time she starts pulling the ends of the thread apart and she gazes into the small crowd, her eyes telling us all 'and as if by magic …'. The byssus is quite stretchy and elastic.

Now she jumps to her feet and bustles to the window where she holds up the thread to show us all how it gleams a bright golden hue. She breaks the thread in two and presents one piece each to Rebecca and me.

The performance complete and the thread of sea-silk made, Chiara glides around the room showing us some of the things she creates. Individuals and organisations around the world commission her to make weavings and embroideries. A group of Nelson enthusiasts have recently been in touch asking Chiara to weave them a pair of sea-silk gloves like those of their hero. She brings out a small square of knitted sea-silk with a fine open weave and lays it in my hand; it is delicate and dainty but I'm still not quite convinced it would fit into a walnut shell. Many of her works are for churches and cathedrals, and she shows us a splendid embroidery of Mary and baby Jesus. There are no price tags, and nothing is for sale. Such a commercial venture is quite against her ethos of working with her one great collaborator, the sea. This is an entirely voluntary endeavour, made possible only by generous donations dropped in the box by the museum's door.

In a wooden frame is a golden embroidered lion, with a fancy tail and its front paw raised. It was made several decades ago by Chiara's grandmother, the woman who taught her how to make sea-silk. Chiara tells us how she believes her family has made sea-silk for 30 generations (by my calculations that is somewhere between 600 and 900 years). Other people in Sant'Antioco tell me that Chiara's grandmother, just like Efisia Murroni, learnt the skills of sea-silk spinning and weaving from Italo Diana.

A row of containers on a stone windowsill are filled with coloured liquids. Chiara picks up a purple jar and swirls it around. This is the infamous dye that is produced from several species of marine mollusc. Murex shells were dredged up from the Mediterranean in their millions and crushed to produce the rich imperial and Tyrian dyes used to colour the robes of ancient Phoenicians and Roman emperors. Chiara shows me a tuft of byssus with a subtle lilac hue. Dyeing sea-silk like this is a technique that has been passed down, so she says, through generations of sea-silk weavers in her family. If this is true then they were probably the only ones doing it: there are no records of sea-silk being tinted with these molluscan dyes, or any other pigments for that matter, besides the lemon-juice treatment.

The one thing Chiara will never reveal about her sea-silk weavings is how exactly she gets the byssus to make them. Now in her fifties, she tells me she has known for 30 years how to extract fibres without damaging living pen shells. Now the shells are protected, this has become a necessity.

The precise details of how she does this remain a carefully guarded secret. She distrusts the biologists who ask to watch and study her at work, convinced they will steal her ideas and open up a new sea-silk industry that will devastate the pen shell population.

All she will say is that there are certain times of year, and certain phases of the moon, when the seabed around Sant'Antioco becomes soft enough to gently pull the pen shells from their resting places. Helped by a local, trusted fisherman she dives down without scuba gear, so she says, and snips off 10 centimetres (four inches) of byssus from each living shell, like giving them a haircut or trimming their nails. Then she pushes each giant shell back into the mud. Is this a genuine technique, or just another part of the mythology she weaves around herself?

As a cool wave of reality ripples into Chiara's world and mingles with her stories, it's hard to know for sure what is actually going on. She can't legally be taking whole pen shells, and her website states that her annual sea-silk harvest is around 600 grams (about 20 ounces). If she only trims their beards she must have to process thousands of shells every year (the full beards from 50 shells will yield only around an ounce of sea-silk). She would then have to leave them alone for long enough to recover, assuming they survived. Maybe there are enough pen shells living in the waters around Sant'Antioco to support a rotational harvest like this without impacting the population; but no one, except perhaps Chiara, really knows if this is the case.

A couple of important questions hover over Chiara's claims of a sustainable byssus harvest. First, whether the shells do indeed survive through the harvesting process and regrow their trimmed beards. Based on what is known about the biology of pen shells and other byssus-making bivalves, there is a good chance that the shells do survive, so long as their internal byssus gland remains intact. If it does, then the pen shells would need to grow whole new byssus fibres to re-root themselves in the seabed. With their ends cut off the fibres lose their sticky pads, but that shouldn't be a major problem. Many bivalves grow new byssus filaments throughout their lives, replacing ones that break off. Some even use them as a way of moving over the seabed, throwing out a line, then hauling it in with retractor muscles and shuffling forwards.

Another unanswered question is how long it takes pen shells to grow new byssus beards, and re-root themselves. Until they do, the shells have to stand up on their own, wedged into the mud and sand without the stretchy anchor securing them in place. If a shell gets knocked over it has no means of righting itself and could choke on the seabed and find itself vulnerable to nibbling predators. That said, if the shells are in sheltered, calm water there is much less risk of them falling over.

Judging by other species, the rate of byssus growth could be reasonably speedy. It only takes a few minutes for a Blue Mussel to make a single new fibre, although they are much shorter than pen shell byssus. Mussels can make up to 50 fibres a day, but they will speed up or slow down production depending on various factors. Fast water currents stimulate mussels to make more fibres, although only within reason (if the water flows too fast the mussels find it impossible to get a grip). The whiff of predators like crabs and starfish is enough to trigger byssus production, presumably because this fixes them more firmly to the seabed, making them difficult to eat.

Poking mussels to simulate an exposed, wave-rattled shore is another way of motivating them to get busy making more byssus. In one study mussels were agitated at different rates of between once every 4.5 and 27 seconds for up to two weeks at a time (this was done by an automated mussel-bothering machine, not a sleepless grad student); the more the mussels were disturbed, the more fibres they made.

Being able to control the rate of byssus production is important, because the process is hard work. Making these fibres uses up a lot of energy and protein, which is why mussels will only make as many fibres as are needed according to the prevailing conditions and risk of attack. It is possible that pen shells respond to Chiara's trimming by ramping up byssus production, diverting energy from other parts of their body to do so. How this affects them isn't known.

It would be easy enough to find out whether pen shells with trimmed beards do indeed re-root themselves in the seabed and how long it might take, but these aren't research topics that anyone has yet pursued. An *ad hoc* experiment did get underway in 2012 when the *Costa Concordia* cruise ship hit a rock and sank off the coast of Italy. While the ship lay on its side at the surface like the beached carcass of a giant white whale, divers surveyed the water beneath and found a nearby seagrass meadow with a population of around 200 pen shells. They decided to move them out of harm's way.

News coverage on the internet shows divers gathering up the shells and stacking them temporarily in plastic crates on the seabed. The plan was to put the shells back in their original location, replanting them in the seabed, once the wreck was salvaged. The outcome of this transferral will help demonstrate whether pen shells can cope with being handled. Chiara tells me how annoyed she is about all this, because she thinks it gives people the idea of pulling up pen shells and making sea-silk. She worries that if the masses blunder in and copy her, it will end in disaster for her beloved pen shells.

The only way I see these giant seashells becoming endangered because of their byssus would be if new markets or appetites arose, if sea-silk became the darling of fashionistas or the substance of some other fetish. If that ever happened then a truly sustainable byssus harvest, one that doesn't lay waste to wild pen shells, would be an unlikely dream. The real world shows us that this sort of thing almost never happens.

Look at the vicuña, a wild relative of alpacas and llamas, which lives on high grassy plains in the Andean mountains. To stay warm, these dainty camelids grow ultra-fine wool that can be spun into a fine and expensive fabric. The Peruvian government set up a labelling system for wool taken from animals that are caught at most every two years, shorn and released unharmed. Of course it is much simpler to simply shoot a vicuña and skin it. Vicuña numbers are recovering but poaching continues, as does the black market in cheaper, uncertified wool. A similar situation would probably unfold if there was ever a market for sea-silk. Luckily so far, though, demand for sea-silk remains negligible.

Chiara is kindling a desire for sea-silk but she is also fiercely protective of its source along the shores of Sant'Antioco. In many ways, she is doing the opposite of philanthropists who came before her, who tried to stimulate the sea-silk industry and help other weavers make a living.

The rarity of sea-silk fibres and the difficulty of obtaining them is a challenge Chiara faces, but at the same time it is the key to her fame and success. She clearly needs to protect the source of these delicate fibres together with the museum and livelihood that rely on them. By retelling folktales and weaving new traditions to fit with the modern world, Chiara is getting caught in the threads of her tapestries and becoming part of the story herself, and in doing so she guarantees the spotlight stays focused on her as the self-styled saviour of a fading custom and craft.

Stepping outside into the bright sunshine I clutch my piece of sea-silk, and ancient stories waft through the museum door behind me. Suddenly it strikes me how bizarre it is to be holding a piece of thread made from fibres that oozed from a mussel. Then again, why is it any stranger than wool that grew on a sheep's back, or silk that was spat out by a caterpillar? It reminds me of the extraordinary brocaded cape woven from the silk of a million golden orb-weaver spiders in Madagascar, and displayed in London at the Victorian and Albert Museum in 2012.

I had seen sea-silk being made and was quite convinced that this stuff really does exist, but there remained one part of the story of sea-silk I wanted to see.

I walk down the hill to a small wharf where fishermen are unloading octopuses, amorphous handfuls of soft white glop, while others tout for business; these days there is more money to be made taking tourists on fishing trips than selling the fish they used to catch themselves. Passing the large boats, kitted out for a day of fishing and feasting, I come to a smaller wooden boat painted blue and filled with ropes, polystyrene buoys and a pair of worn oars. A small fish, perhaps a goby, stares at me from the transom, dead but only recently. The skipper helps me clamber on board and I wonder if we will both be rowing. Then a little engine hidden in the stern kicks into life and we chug across the flat lagoon to a spot just offshore. For an anchor, he pushes a wooden pole through a hole in the hull, pinning us to the shallow seabed below while I scramble over the high gunwale and into the cool water.

Paddling around, snorkel in my mouth and eyes down, I catch my first glimpse of the Noble Pen Shells, nestled in a lush garden of seagrass and seaweeds. The shells look frilly and soft on the outside, but I discover they are firm to the touch as I reach down through the shallow water and gently tap one with my fingernail. When I do, the shell twitches, pulls in its white and black-flecked mantle and slowly shuts. It closes its mouth into a puckered semi-circle facing the surface above.

Living among the pen shells are plenty of other creatures. Tiny green fish dart constantly around me. A bright red starfish is splayed out on one shell, and Peacock Worms stick their heads from thin tubes, each one unfurling a crown of feathery tentacles. I spy a bubble snail, a type of sacoglossan sea slug, sliding across a pen shell; it carries its own fragile shell on its back, like a precious marble clenched between two folds of lime-green mantle, a precise colour match for the *Caulerpa* seaweed it lives in.

I sneak up as slowly and quietly as I can on a few pen shells and peer inside to check for hiding crustaceans. For a long time, people have known about (and often become slightly obsessed with) the tiny creatures that live inside pen shells.

Known generally as pea crabs, they are commonly depicted as sentinels, watching over the blind molluscs and alerting them when trouble or food is near. Pliny the Elder described the pea crabs as signalling to the pen shells with a gentle nip whenever a little fish wandered in; the shell slams shut and then both mollusc and crab tuck into a shared dinner. More recent studies reveal there are two crustacean species associated with pen shells – a crab called *Nepinnotheres pinnothere* and a shrimp, *Pontonia pinnophylax* – but they aren't security guards or hunting partners; the shell interior simply provides them with a safe refuge. The crabs eat the same planktonic food as the filter-feeding shells and the shrimp scrape food particles from the surface of the molluscs' gills and snack on their pseudofaeces.

People have done strange things with pinna pea crabs. There are ancient recipes listing pea crabs as an ingredient for soup. They have also been a source of moral guidance, with the belief that we could all be a bit more selfless and cooperative like them. An ancient Greek book from the second century called *The Interpretation of Dreams* informs couples that they will have a long and happy marriage if they dream about the pinna shells and pea crabs that live so harmoniously together. Strange dreams indeed.

I scoot from pinna to pinna, but none of them seem to be occupied by little crabs or shrimp. Inside a dead shell, gaping and still, there is a dark shadow of a fish lurking. The basilisk blenny slowly retreats like a shadowy face pulling back from the window of an abandoned house, not wanting you to know that you are being watched.

Most of the Noble Pen Shells I see are on the small side, about as wide as my outstretched hand. They are all young ones, not yet fully grown. It means this particular spot is an important nursery for the population, and a good indication that all is well for the pen shells of Sant'Antioco. I can't tell for sure without spending days and weeks swimming around the entire island counting shells as I go and then ideally coming back some time later to see if things have changed. But the presence of juveniles is a sure sign that adults are nearby and they've been successfully breeding. These are probably not shells that Chiara will harvest, because they are too close to town, and to the gaze of prying eyes. Popping my head up above the surface, I see a coach-load of tourists drive past along the seafront, a few hundred metres away.

Not much is known about pen shells and their current status in the Mediterranean, following protection more than 20 years ago. A few scientific studies have mapped out their distribution and sizes, and there are signs of recovery and healthy populations. Their seagrass habitats are certainly under pressure still, in particular from rising sea temperatures, but pen shells do live elsewhere, too, in sandy, muddy environments that are far less threatened. To some extent the pen shells' protection is a precautionary measure, a proactive step to make sure they don't dwindle as they so easily could, rather than waiting for catastrophe to strike, by which time it might already be too late.

Back down below me, the Noble Pen Shells seem to shift and glide across the seabed but in fact it is the grasses and weeds that flutter in the breezy current around them, while the shells stay put. They are wedged firmly in place up to their middles in the soft sediment, anchored by their unseen byssal threads.

There is no doubt that sea-silk continues to enchant people, especially when they are regaled with worn-out fables told as if they were still true today. Surely, though, there are wonders enough to be had in the reality of these giant shells with golden beards. We can marvel at the tiny crabs that cohabit with the living shells, and the octopuses and fishes that move in when they die; we can ponder the strange mystery of who it was who first thought to tease out a pen shell's fibres and spin them into silk; we can contemplate the spelling mistake made centuries ago that led to a deep-rooted case of mistaken identity; and we can admire the intricate embroideries made by the artisans of the more recent past and present.

Giuseppina and Assuntina Pes will keep working on their weavings but for the most part they will use alternative fibres, not sea-silk. Chiara Vigo will continue to run her museum, tell her stories and venture to the shore to gather more byssus when no one is watching.

The Noble Pen Shell is a rare thing indeed. It is a sea creature with something to offer but isn't, for once, being plundered to meet human needs and desires. So it can only be a good thing that newly woven sea-silk remains an obscure, curious thread that gleams now and then on just one tiny island.

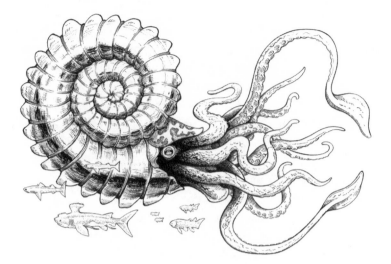

CHAPTER SEVEN

Flight of the Argonauts

I've never seen a living argonaut. Few people have. In a rare sighting in October 2012, fishermen accidentally caught a female argonaut while hunting for squid a few miles off the coast of Los Angeles. They brought the strange creature back to shore and gave it to a local aquarium. It was unusual for this tropical species to show up in temperate Californian waters. Staff at the Cabrillo Marine Aquarium assumed that she had been carried on a current sweeping up from the south and carefully placed her in a warm-water tank. For some time, the exhausted animal lay helplessly at the bottom and the aquarium keepers feared the worst. Then one of them thought to give her a helping hand towards the water surface. After that, the argonaut perked up, and started swimming around her captive home; she eventually began to eat, grabbing morsels of fish and shrimp offered to her.

A video posted online shows the captive Californian argonaut. Hovering in the water, her shell is iridescent with a bronzy-silver gleam and for the first few seconds it's difficult to make out the animal inside. Then all of a sudden she pops out, revealing herself to be a delicate, shiny little octopus. She pulls out her eight arms, grabs hold of her shell and deftly spins it round before climbing back inside.

Argonauts are the only octopuses that live inside a shell. All the other members of the order *Octopoda*, around 300 in total, have embraced a soft, naked life. Now and then you might spot a common octopus peeping out from inside an empty clam shell. A video clip went viral a few years ago of an octopus in Indonesia picking up half a coconut shell and strutting off across the seabed, using its arms as legs. When it comes to full-time shell-living, though, it's just the four members of the genus *Argonauta*: the Greater, Rough-keeled, Brown and Tuberculated Argonauts. They all look quite alike, with pale and thin shells, covered in ridges and rows of nodules. Depending on the species, their shells can be between five and thirty centimetres (two and twelve inches) across, while the animals inside are considerably smaller. Throughout their lives they cruise the upper highways of tropical and subtropical seas, way above the heads of their octopoid relatives, which mostly live close to the seabed, lolloping and swimming along but rarely venturing too far up into open water.

After a week of life in captivity at the Cabrillo aquarium, the argonaut gave everyone a big surprise. She was joined in her tank by thousands of tiny argonauts. It turns out she had been carrying fertilised eggs, and now they were starting to hatch.

It was all hands on deck as helpers were drafted in to count the new arrivals. Clerical staff were brought out from behind their desks, and visiting schoolkids were given a taste of scientific research. Over a course of a few days, the argonaut released a total of 22,272 minute hatchlings, each one only a millimetre across. Other videos, this time shot down a microscope, show some of the new argonauts. The twitching oval blobs are mostly transparent, with two big, dark eyes and a covering of spots that expand and contract; one minute they are patterned like a giraffe, the next they are peppered with tiny black dots. The flickering colours are made by chromatophores, cells embedded in the mantle that are filled with pigment granules and are concealed or revealed by minute muscles relaxing or contracting. The infant argonaut grapples with zooplankton and uses its little arms to shovel them into its mouth; it's the first time such a tiny argonaut has been caught on camera tucking into its food.

Sadly, though, the Californian argonaut and her plentiful offspring didn't survive more than a few weeks in captivity. The aquarium keepers couldn't easily have returned her to the sea because the warm water current that delivered her to California had stopped and they were a long way from her normal tropical habitat. At around the same time, empty argonaut shells were found washed up on nearby beaches, suggesting there had been some sort of mass stranding. Even if the captive argonaut had been left at sea she might not have survived. At least this nomad had helped researchers gain new insights into these most enigmatic creatures.

People have known about and puzzled over argonauts for millennia. Two questions have confounded many great minds: what purpose does the argonaut's shell serve, and where do their shells come from?

The name 'argonaut' stems from Greek mythology, and the band of heroes – the original Argonauts – who sailed on the ship *Argo* with Jason in search of the Golden Fleece. It was the Greek philosopher Aristotle who first wrote about their molluscan counterparts. He suggested they use their shells as boats to float on the surface of the sea, with their arms as oars to row themselves along, or two arms flattened and hoisted up as sails. The story was passed on and retold for centuries by naturalists, and writers who professed to have seen this strange scene for themselves. The sailing octopuses appear in Jules Verne's 1870 novel *Twenty Thousand Leagues Under the Sea*. While held captive aboard Captain Nemo's submarine, the *Nautilus*, marine biologist Professor Aronnax ponders the peculiar sight of hundreds of argonauts sailing across the waves, all holding their arms in the air like flapping ears.

An alternative common name for argonauts is the paper nautilus, because their light, papery shells look a little like those of the chambered nautilus. As this name suggests, nautiluses have shells that are divided into chambers (argonaut shells, by contrast, have no inner chambers). As they grow, expanding their shells from the open end, nautiluses inch their body forwards, and periodically seal a chamber off behind them. A tube running between the chambers, called the siphuncle, then empties liquid from the new chamber by osmosis, and gases diffuse in. Nautiluses can adjust the fluid levels inside their shells, like a submarine's ballast tanks, controlling their buoyancy and reducing the energy demands of active swimming. Like other cephalopods, nautiluses swim by jet propulsion in a two-stroke system: water is sucked inside the shell, then squeezed out through a funnel. Shifting the position of the funnel controls their direction to some extent; nautiluses swim hesitantly forwards but can scoot away backwards at much greater speed. When they feel threatened, they can withdraw inside their shells, and shut the opening with a leathery trapdoor called a hood.

On the inside nautilus shells are lined with mother-of-pearl, giving them their other common name, the pearly nautilus. On the outside, they're decorated with ginger tiger stripes across the top, with some that fade to white underneath, as if on being dipped in the sea their markings had started washing off. There are four recognised species in the *Nautilus* genus, including the Belly-button Nautilus and the White-patch Nautilus. Two other species were shuffled across into a new genus, *Allonautilus*, because when living specimens finally showed up a few years ago they were thought to be rather too different from the rest. All of them have around 90 slim tentacles – the most of any living cephalopod – making them look like they're eating a mouthful of spaghetti. They occupy tracts of deep, tropical waters, in the Indian and Pacific Oceans, and are rarely seen alive. When they die their empty shells bob to the surface and can drift to distant shores.

Empty shells were all people knew of nautiluses for a long time. Collectors adored their shininess and elegant whorls, and naturalists were desperate to get their hands on a complete specimen, soft parts and all. Paper nautilus shells, on the other hand, did occasionally show up with something living inside them, but this didn't stop naturalists arguing over the identity of these little creatures.

On an ill-fated 1816 expedition to find the source of the River Congo, British naturalist John Cranch was fishing for specimens from the Gulf of Guinea off West Africa when he found several argonaut shells, complete with living occupants. They survived on board in a bucket of seawater for several days while Cranch observed them. He saw they could come all the way out of their shells, if they wanted to, and otherwise looked and behaved like octopuses: they had suckers that stuck to the side of the bucket, they swam around using a jet of water and their skin changed colour.

All of this was later reported by William Leach, curator of zoology at the British Museum, because Cranch himself died of fever, along with most of the crew, and never made it back from Africa. In honour of his lost friend, Leach named the argonaut species *Ocythoe cranchi*, but this was applied only to the soft animals, not the shells they were found in. Many eminent naturalists believed the octopuses didn't belong with the shell but had killed and devoured the original occupant before seizing their vessel and sailing off. In their minds the octopuses were nothing but parasites.

Carl Linnaeus had named the shells *Argonauta argo* back in 1758, in the tenth edition of his book *Systema Naturae*, and in 1814 Constantine Samuel Rafinesque assigned the name *Ocythoe antiquorum* to the allegedly parasitic animals that were often found inside. John Cranch's was a new species of parasitic octopus.

For a long time, a living specimen of the shell-making argonaut itself remained unknown. Presumably they were lurking down in the depths somewhere; perhaps they were some other kind of nautilus. The fact that none had been found wasn't seen as a major problem, though; after all, chambered nautiluses were very rarely seen alive, but their empty shells were fairly common.

In 1828, English naturalist William Broderip reported in the *Zoological Journal* that a French collector in Marseille claimed to have found a *real* argonaut, not a hitch-hiking *Ocythoe* octopus. I can sense Broderip's eyebrows twitching as he wrote this, and he stayed on the fence, pointing out that much remained to be known before coming down firmly on either side. But he still took a punt that in the long run the octopuses would probably be revealed as pirates, and not the industrious shipwrights of what he called 'fairy boats'.

The idea of octopuses sailing around in stolen shells may sound like a fanciful Just So Story, but there are some even stranger ideas floating around that have made scientists stop and think. Rather than snatching shells from living species, maybe argonauts hijacked them from far more ancient creatures?

The small collection of living nautilus species is all that remains of an immense cephalopod dynasty. In modern seas, the most common cephalopods are the ones with no external shells, the octopuses, squid and cuttlefish. But in times gone by it was the shelled cephalopods that reigned supreme. Masses of animals that looked a lot like nautiluses romped through the oceans for hundreds of millions of years. Within that group, the most abundant and diverse of them all were the ammonites, and there were some that looked so eerily similar to argonauts, you might be persuaded they were cast from the same mould. By the late nineteenth century, a distinctly offbeat idea had come to light. What if naked octopuses originally borrowed or stole shells from ammonites? Did argonauts learn how to make shells by copying their ancient relatives?

This theory was first proposed in 1888 by German geologist Gustav Steinmann; it was revisited in 1923 by Swiss palaeontologist Adolf Naef, then again in the 1990s by Zeev Lewy from the Geological Society of Israel. They all imagined the ancestors of modern argonauts to have started out hiding inside empty ammonite shells. Then the argonauts somehow evolved the ability to fix up their borrowed shells, to mend holes and cracks. As the argonauts got better and better at repairing shells they eventually no longer needed a template at all, and could merrily continue shell-making without having to find an ammonite shell first.

Lewy went a step further, proposing that argonauts were in fact scavengers of recently dead ammonites, which he rather charmingly referred to as 'post-necrotic floaters'. In other words the ammonite shell, complete with dead animal inside, floated to the sea surface and drifted around for a while. Lewy suggested that naked argonaut ancestors laid their eggs inside these post-necrotic floaters, leaving the new hatchlings to slowly eat their dead hosts and ultimately occupy the vacated shell.

To find out if there is any truth in these ideas and see if there is a link between argonauts and ammonites, we should jump back in time half a billion years to see where this all began. Down at the base of the cephalopod evolutionary tree sits a little creature that lived towards the end of the

Cambrian. It was about the size of a pinky toe and wore a slender and slightly bent shell like a wizard's hat. Charles Doolittle Walcott, of Burgess Shale fame, was the first to describe fossils of these animals (though they were from later deposits), and he named them *Plectronoceras*.

Plectronoceras is the oldest undisputed cephalopod (strange creatures called *Nectocaris* from the Burgess Shale itself could be cephalopods, although not everyone agrees on that). Their shells were divided into chambers, like nautiluses, and they may have spent much of their lives skipping across the seabed or wafting through shallow seas as part of the plankton. Following on from these modest drifters there were far more impressive, not to mention scarier, cephalopods to come.

Starting around 485 million years ago, the Ordovician was the next major stage in Earth's history. The planet was a strange place compared to the way things are now. Temperatures were much higher, as were carbon dioxide levels, and most of the landmasses were clumped together into a massive super-continent, Gondwana, but nothing much lived there. Life was still largely confined to the oceans, where there was a mixture of living things that we could recognise today, plus a range of other, bizarre creatures.

Trilobites scuttled across the seabed; bivalves and brachiopods stayed put as they sifted the water for food; gastropods ambled past fronds of red and green seaweeds and colonies of coral. Above the seabed, early chordates called conodonts wriggled their eel-like bodies and gnawed at their food with the sharpest teeth that ever evolved; floating through the water were colonial creatures called graptolites that looked like delicate, saw-toothed tuning forks. For all of these creatures, one of the most dangerous things they were likely to encounter in Ordovician seas was an enormous shelled cephalopod.

The unassuming Cambrian cephalopod lineage went on to flourish in the Ordovician. They evolved into masses of new groups; some were tightly coiled, others had shells as straight as pencils. Incomplete remains have been found of gigantic straight shells from a creature named *Cameroceras*. Estimates of their full size range up to an astonishing 10 metres (more than 30 feet), as long as a double-decker London bus. These were formidable beasts, like a primeval apparition of a Colossal Squid, the main difference being that these ancient creatures lived inside the longest seashells ever to exist.

It's generally thought that *Cameroceras* may have spent a good deal of time resting close to the seabed, pulling itself along with a cluster of arms and scooping prey into its mouth. Other straight-shelled cephalopods would have hung in the water with their heads down, grabbing prey from the bottom. Some evolved counterweights at the ends of their long shells and swam horizontally. Like giant spears, they could have shot through the oceans in pursuit of prey. Whichever way you look at it, the Ordovician saw the rise of the cephalopods.

Towards the end of this period, the super-continent Gondwana drifted towards the South Pole, giant ice sheets spread across the land and Earth fell into a very deep ice age. Sea levels dropped, and continental shelves were drained of their shallow seas, depriving much marine life of its habitat and triggering a mass extinction. Over half of all marine invertebrates were wiped out, but cephalopods were among the survivors.

For tens of millions of years, cephalopods waxed and waned many times. Throughout the Silurian and into the Devonian periods, they went through repeated pulses of decline but always picked themselves up and carried on, recovering their abundance and diversity. It was in the early part of the Devonian, around 400 million years ago, that a series of important new branches sprouted in the cephalopod evolutionary tree. There were the Nautilida or nautilids that led on to the modern nautiluses. The coleoids showed up too, which eventually gave rise to the living octopuses, cuttlefish and squid. The third major lineage of cephalopods to emerge in the Devonian went on to produce some of the most supreme sea creatures of all time: the ammonites.

Chronoscopes and thunderstones

If you fancy getting your hands on your very own ancient, extinct creature I'd recommend looking for an ammonite. Fossil ammonites are hugely abundant and widespread, not to mention very lovely objects. I have several ammonites that were found and given to me by Kate, my geologist sister, who knows only too well my soft spot for interesting things from the sea. My favourite in this little collection is an intricate, tightly coiled shell covered in delicate ridges, and just the right size to cover up with my thumb. It got trapped in a layer of black silty mud that eventually turned to mudstone and became part of the crumbly cliffs of Kimmeridge Bay on England's south coast. This animal swam through the seas 150 million years ago and now sits on my desk, where from time to time it helps to straighten out my sense of perspective on the world, and of time passing.

Because they're so common and easy to find, fossil ammonites have been wending their way into human lives for thousands of years, sometimes without people even realising. Walk through the Grand Arcade shopping centre in my home town of Cambridge, England and look down, and you'll spot ancient spirals in the polished limestone tiles beneath your feet. Long before anyone knew their true origins, and way before they began appearing in shop floors, people across the globe found these strange coiling stones and wondered what they were.

In Europe, fossil ammonites were often called snakestones, with accompanying legends to explain how they were made. Often it was a story about a saint, who went around turning real snakes into stone then hurling them off cliffs.

Snakestones were widely believed to cure snakebites and all sorts of other conditions, from human impotence to cramp in cows.

Ancient Romans believed they would see into the future if they slept with a golden ammonite under their pillow. The Blackfoot people of North America thought ammonites looked like sleeping bison and called them buffalo stones; finding one before a journey was a good omen. Black ammonites from the Gandaki River in the Himalayas are called shaligrams. They are worshipped in monasteries and temples as manifestations of the Hindu god Vishnu, and people on their deathbeds drink water steeped in these sacred stones to free them of their sins.

Similar beliefs surround belemnites. These extinct relatives of the ammonites were coleoids, along with octopuses and squid, and while they were quite squiddy in their external appearance they had an internal, bullet-shaped shell. Fossil belemnite shells, known as thunderstones, were thought to be created when thunderbolts struck the ground, and they too were used as a cure for snakebites, as well as protecting a house from getting hit by lightning when they were placed on a windowsill. In Swedish folklore thunderstones held strong magical powers that guarded against evil; they were thought to be candlesticks used by supernatural creatures called *vättar* that live under the floorboards and cause trouble if the house isn't kept tidy (in some versions of the story they are distant relatives of Santa Claus). In eighteenth-century England, fossil belemnites were ground down and used as an ointment for horses with sore eyes. In Scotland, the traditional name for them was botstone; people would drop one in a horse's water trough to treat a case of worms.

Bountiful fossil ammonites have also been put to practical use. In Victorian Britain, they were dug up and used to make the world's first artificial fertiliser. As urban populations grew and more mouths needed feeding, scientists discovered that phosphate was a key ingredient for growing better crops. Expensive bird droppings, rich in phosphate and known as guano, were imported from Peru at substantial cost. Animal bones from knacker's yards, shavings from bone-handled knife factories, mummified Egyptian cats and allegedly even human remains from European battlefields were all ground down and sprinkled onto arable fields. Then a source of phosphate was found much closer to home. Buried deposits of fossilised bones, mixed in with assortments of ancient animal teeth, claws, shells and the droppings of extinct marine reptiles were found to be an excellent source of phosphate. The concoctions came to be known as coprolites, from Greek words for dung and stone, even though only some of it was actually petrified poo; everything else technically should be referred to as pseudo-coprolites or better still, phosphatic nodules. In the mix were ammonites; after they died, the calcium carbonate in their shells was replaced with calcium phosphate from seawater.

A shallow Cretaceous sea that used to cover south-east England winnowed fossil ammonites from older rocks and swept them into dense piles. It was these ancient relics that triggered a coprolite mining rush and saw open-cast mines appear across the country. Great fortunes were made in digging up and processing coprolites, in particular around the city of Cambridge, where almost all of Britain's raw phosphate came from.

The Sedgwick Museum of Earth Sciences in Cambridge has display cases filled with coprolites. Many of them were found by Harry Seeley, an assistant to Cambridge's professor of geology in the mid-nineteenth century, Adam Sedgwick. Throughout the 1860s, Seeley paid regular visits to the nearby coprolite pits where he picked through the washing tanks to see what interesting and unusual specimens were turning up. On display today at the museum are grey and black ammonites, as well as bivalves and gastropods.

Besides the few specimens liberated by Seeley, estimates suggest another two million tonnes of phosphate-rich fossils were dug up and loaded onto horse-drawn carts, steam trains and barges and taken away to be crushed in windmills. Sulphuric acid was then added to the powder to make 'superphosphate', which was sold for half the price of Peruvian bird droppings and was exported across the globe. Until cheaper sources of rock phosphate were found in the 1880s and coprolite production fell, arable crops from Russia to Australia were grown with the aid of some very old seashells.

Fossil ammonites have left another, more lasting legacy in the human world. Two hundred years ago, British engineer William Smith was the first person to realise that fossils, and in particular ammonites, were time capsules that declare the age of rocks. His job involved travelling the country, digging a new network of canals. He noticed that as his men dug deeper the rocks changed, and so did the fossils inside them. He gathered together a fine collection of fossils, including many ammonites, and used them to prove that rocks are deposited in flat layers like pancakes; later those flat rocks can become squashed, tilted and folded as the Earth's crust shifts.

Several features of ammonites made them extremely useful to Smith as he probed geological formations. Not only were their fossils immensely abundant and easy to find, but there were also thousands of ammonite species (many can be identified from intricate patterns like fingerprints, called sutures, etched across their fossilised shells; these were the junctions between the internal chamber walls and outer shell, with the lines revealed when sand and mud filled an empty shell, then formed an internal mould). Individual species also tended to be quite short-lived, appearing and then going extinct in a geological heartbeat, sometimes just a few hundred thousand years. This means that if the same ammonite species is found in different locations, the rocks they lie in must be roughly the same age. This is the basis of a powerful geological technique known as biostratigraphy. With their cosmopolitan ranges, ammonites assist geologists in linking rock formations on opposite sides of the planet. The same species have been found in Chile, Australia, Europe, Madagascar, China and Antarctica.

By matching the ages and types of rocks in different places, Smith drew an enormous map, two metres (more than six feet) tall, showing in fine detail the geology of England, Wales and part of Scotland. With different colours for different rock formations, he produced a rainbow view of the British Isles that had never been seen before. The map and Smith's findings played an important part in the emerging science of geology, helping to advance theories of how rocks are formed over millions of years.

You say ammonite, I say ammonoid

A confusing thing about ammonites is that, technically, rather a lot of them should in fact be referred to as something else. The lineage that ammonites belong to – the ammonoids – split from the rest of the cephalopods in the Devonian around 400 million years ago. The true ammonites showed up more than 200 million years later, in the Early Jurassic. Before then, dozens of other ammonoid groups came and went. People usually refer to them all as ammonites, but in fact they were different, closely related animals.

In Palaeozoic seas, from the Devonian onwards, the dominant ammonoids were the goniatites, most of them with small, compact spiralling shells. They thrived until 252 million years ago when a crisis hit the living planet, one like none that had come before. The End-Permian mass extinction, also known as the 'great dying', was probably triggered by a combination of colossal volcanic eruptions, the bubbling up of methane from the deep sea and subsequent runaway global warming. It wiped out 70 per cent of life on land and 96 per cent of ocean-going species, including the last of the trilobites.

Even though the goniatites went extinct, the ammonoid lineage survived into the Triassic. The oceans filled with the next major ammonoid group, the ceratites. They were quite short-lived, with a reign that lasted only 50 million years or so. Then one final, grand assembly of ammonoids took centre stage. From the early Jurassic onwards, the oceans were teeming with ammonites.

Even though their fossils are incredibly abundant, the ammonites and their relatives remain deeply mysterious creatures, and many of their secrets remain locked in the past. Apart from their shells, we don't know what ammonites looked like. So far, not a single fossil ammonite has been found with its soft body preserved. Did they have eight arms like octopuses? Eight arms and two tentacles like squid and cuttlefish? Or did they have dozens of noodly appendages like nautiluses? We don't know.

One thing we do know is that they probably swam around by jet propulsion. A notch in the opening of ammonite shells hints that they had a fleshy funnel, like living cephalopods. It's mind-numbing to imagine the biggest known ammonite, *Parapuzosia,* pulsing its way through the seas – fossils of their shells are two metres (six and a half feet) in diameter. Experts think the living creature could have been three metres across, and weighed a tonne and a half or more. If giants drove around in monster trucks, these shells would be their wheels.

There were plenty of other strange sights in the oceans during the reign of the ammonites. On the whole, their shells were sculpted into neat spirals; they occupy just a small corner of David Raup's museum of all possible shells. Some ammonites, though, did things completely differently.

Helioceras was an ammonite with a tall, helical shell covered in spikes that looked like a dangerous helter-skelter. They would have hung with their heads down, and a gentle puff of water from their funnel would have sent them into a spin. Perhaps they pirouetted up and down through the seas like corkscrews. *Nipponites* was another strange ammonite. It had a meandering shell, tangled up in knots, similar in appearance to (but much bigger than) the microsnails that live today in the chalk hills of Borneo.

Something else we don't know about ammonites is what they ate. Rare fossils have been found with what could be their stomach contents, including little creeping crustaceans called ostracods and flower-like relatives of sea urchins called crinoids, as well as other ammonites. But not everyone agrees that these definitely were the ammonites' last meals. What is clear, though, is that other animals were eating ammonites. They were not the highest-ranking predators in the oceans, as their Ordovician ancestors were. The hunters had evolved into the hunted.

Fossil ammonites have been found with smooth, round holes in them and some experts think these are scars left by limpets that latched on after the ammonites died. Further analyses, however, point towards a more brutal endgame.

Jurassic ammonites shared the oceans with plenty of scary beasts, including dolphin-like reptiles, the ichthyosaurs, followed later by mosasaurs. These were terrifying marine lizards, up to 20 metres (65 feet) long with huge snapping jaws packed with teeth that just happen to match the size and spacing of the holes in many ammonite shells. Rather than limpet scars, a more likely explanation is that the holes are indeed tooth-marks. There seems to be no obvious reason why limpets would line themselves up, time and again, into the same V-shaped arrangements.

One ammonite has been found with punctures in two sizes: a perfect fit for adult and juvenile mosasaur teeth. Was an adult mosasaur teaching its offspring how to hunt? Or did it sneak up on a youngster and steal its dinner? Either way, it wasn't good news for the ammonite.

Meanwhile, as giant swimming reptiles were chasing after ammonites, new threats to everything in the oceans were approaching. Soon the reign of the shelled cephalopods would come to an end, leaving one final, big question: why are there no ammonites around today?

Ammonites well and truly hogged the cephalopod limelight throughout the Mesozoic. Meanwhile, in the background, another group of shelled cephalopods were quietly getting on with things. These were the nautilids. From the outside, they looked a lot like ammonites but compared with their more famous cousins, they lived in much smaller populations and there were not nearly as many species.

Side by side, the ammonites and the nautilids survived multiple mass extinction events, and kept going until 65.5 million years ago. Then, at the end of the Cretaceous, a mass extinction came along that only one of these two groups would survive.

This is probably the most famous mass extinction of all, because on land it saw the end of the non-avian dinosaurs. It also hit the oceans hard: only one in five marine species pulled through into the Tertiary, and I certainly would have put my money on ammonites being among the survivors, rather than nautilids. There were far more of them, and they were more widespread, two factors that normally create a buffer against extinction. Even so, it was the ammonites that bade farewell to the planet while the nautilids persisted, giving rise eventually to the chambered nautiluses. And for a long time, palaeontologists have wondered why.

To pin down the causes of extinction is difficult enough in the present day. Even when biologists can tiptoe up to endangered species, watch them and test out ideas of why they are in trouble, it can still be a great challenge to decipher the real issues (and even harder to do something about them). Imagine, then, how much more difficult it is when the species in question are already long gone, leaving behind only traces of themselves in rocks. All we have are theories. Researchers have scrutinised the ammonites, then the nautilids and details of the mass extinction, hunting for explanations of what happened and what went wrong for the ammonites.

The longest standing theories about why ammonites lost out are linked to the way they are born. Hatchling ammonites, known as ammonitella, were tiny. We know this because, if you look carefully, you can see the smooth, inner whorls of a fossilised ammonite shell that grew in predictable conditions while it was still inside its egg, feeding off yolk. As soon as it hatched and had to fend for itself in the erratic outside world, new shell layers became irregular. For ammonites, those uneven whorls began to appear when the shell was only one millimetre across. Young nautilids, on the other hand, were around ten times bigger when they hatched. It's thought that at a tender age, these two groups were doing very different things. Ammonites were drifting through the water, as part of the plankton, while nautilids probably stuck closer to the seabed.

This difference may not have mattered too much when the going was good, but it could have been the downfall of ammonites when things got stressful. The exact cause of this game-changing mass extinction is still hotly debated. The fossil record shows that leading up to it, the great ammonite lineage was already in decline, with many genera going extinct. Falling sea levels, which dropped by as much as 150 metres (500 feet) in one million years, may have had something to do with it.

Then, the infamous asteroid, Chicxulub, slammed into Mexico's Yucatán Peninsula, casting dust clouds across the Earth and triggering a long, dark winter. Many experts think this alone explains the extinctions, while others argue that massive volcanic activity in India also had its part to play in the downfall of life on Earth. Today, the Deccan Traps in central India consist of a layer of solid basalt, two kilometres deep and half a million square kilometres in area, which gives an idea of just how enormous these volcanic eruptions and lava flows were. They would have spewed carbon dioxide and sulphur dioxide into the atmosphere, contributing to the planet-wide changes.

Sulphurous gases in the atmosphere would have combined with water and fallen in showers of acid rain; this would have turned shallow seas more acidic and could have made life distinctly uncomfortable for planktonic species, including young ammonites, floating around inside chalky skeletons. By contrast, the next generation of nautilids were tucked up safely down in the deep sea, out of reach of the worst effects of these corrosive waters.

Diet may also have had a part to play in the ammonites' demise. In 2011, Isabelle Kruta and colleagues conducted a detailed three-dimensional scan of an ammonite called *Baculites*. She found what she thinks are remains of its last meal, including the planktonic shell of a gastropod larva. Other experts contend that we can't be sure if this plankton really was food or just a passer-by that got caught in the same rock. But if ammonites did have a microscopic diet, then a collapse of planktonic populations – triggered by corrosive, warming waters during the extinction event – could have left adult ammonites starving.

As for the nautilid diet, their living descendants provide clues as to what they ate. During the day, chambered nautiluses stay hundreds of metres beneath the waves, then rise up at night into shallow coral reefs where they scavenge for the dead. And being seriously short-sighted, chambered nautiluses sniff rather than see their food. They have a pair of sensitive pits, called rhinophores, that help them pick up the whiff of a decomposing body from at least 10 metres (33 feet) away and track the odour plume in three dimensions through the water; scale that ability up to a human standing at the starting blocks of a 100-metre running track, and they could sniff a ripe Brie sandwich being eaten at the finish line. Ancient nautilids may have had a similar habit of smelling and groping their way towards dead food scraps, and it could have made them more resilient to changes in the water around them. Down in the deep, there would have still been plenty to nourish animals that weren't too fussy about their food.

Recently, a new piece was added to the ammonite puzzle when Neil Landman, from the American Museum of Natural History in New York, pondered the importance of geography. He mapped out the global distribution of ammonites that lived towards the end of the Cretaceous, including a handful that survived the mass extinction – for a while, at least. The species that were swiftly snuffed out were ones that had relatively small ranges. By the same token, ammonites that hung on for a while longer generally occupied a wider sweep of the planet. It makes sense that species with smaller ranges are often more vulnerable to extinction. They have all their eggs in one basket, geographically speaking, and are more likely to get wiped out in one go, perhaps by a random event. Imagine a species of dung-eating insect living only in a single cowpat, and what happens if a cow happens to tread on that very turd.

Landman and his colleagues put their findings forward as good evidence that ammonites with a wider range were initially protected, although in the long run it was no guarantee of survival. Ultimately, all the ammonites went extinct (and no palaeontologist truly believes that ammonites could still be out there, somewhere, hiding in the deep). The dying ammonites left the nautilids alone to continue the ancestral line of shelled cephalopods, after almost 400 million years in the sea.

Having followed the rise and fall of the ammonites, let's return to the question of ammonites and argonauts. Could argonauts have learned their shell-making skills from these long-lost ancestors? Nice idea, but there is a fundamental flaw – ammonites and argonauts probably didn't exist at the same time.

We know of 10 extinct argonaut species from the fossilised remains of their delicate shells. The oldest is *Obinautilus* from the Oligocene around 29 million years ago, although some palaeontologists consider this to be a nautilid, which leaves the oldest fossil argonaut at a youthful 12 million years old. Meanwhile, the last known ammonites, as we've just seen, went extinct shortly after the mass extinction at the end of the Cretaceous, around 65 million years ago. More fossils could still be found to fill this gap but, as things stand, it looks highly likely that argonauts never actually encountered any living ammonites, let alone began copying their shells, and it's now widely agreed that this almost certainly didn't happen.

There really is only one plausible explanation for why argonauts have shells that resemble extinct ammonites: it is simply a striking case of convergent evolution. They look so alike because each evolved under the same selective pressure – to be streamlined underwater. Studies have shown that the ridges and ribs on argonaut shells reduce drag while they swim through water, stabilising them and limiting the amount they rock from side to side while they propel themselves along. The same thing would have applied to ammonites too, all those millions of years ago.

Even if argonauts didn't model their shells on ammonites, the question of whether they parasitise some other creatures or make their own shells still needed to be answered. Back in the nineteenth century, argonauts commanded a huge amount of attention and discussion. Scores of scientific papers were written. A few rare preserved specimens of shells with their baffling occupants were passed around. Lord Byron even wrote about them in his poem *The Island*. A host of illustrious scientists held strong views on the argonaut debate, with Richard Owen, Jean-Baptiste Lamarck, Joseph Banks and Georges Cuvier among them. Not everybody was taken in by the stories of octopuses sailing around in stolen boats, and many argued that *Argonauta* and *Ocythoe* should be united as a single species, shell and shell-maker in one. Italian naturalist Giuseppe Saverio Poli examined young octopuses under a microscope and saw they were encased in a little shell, thus proving – he was convinced – that they were not parasites.

In the end, the issue was resolved by a now largely forgotten pioneer of marine biology, who devoted herself through the 1830s to uncovering the truth about these strange animals. Her story of the argonauts follows a meandering journey, taking in a princess's wedding dress and a groundbreaking piece of technology along the way, and ending in the solution to the contentious puzzle of how the argonaut got its shell.

The lady and the argonauts

Jeanne Villepreux was born in 1794, a long way from the sea. She grew up in Juillac, a village in rural south-west France, the eldest child of Jeanne and Pierre. Not a lot is known about her upbringing, but her family seemed to be reasonably well off. Jeanne's father was noted in local records as a shoemaker, shopkeeper, landlord and Juillac's first policeman. When Jeanne was eleven her mother died, and her father remarried. It's not known how well Jeanne got on with her step-mother, who was half her father's age, but she stayed at home until she was 17 before setting off for a new life in Paris.

Chaperoned by her cousin and a herd of cows, Jeanne walked almost 300 miles to the capital. It should have taken around two weeks but the journey was interrupted, so it seems, when her cousin assaulted her and Jeanne sought refuge in a convent in Orléans. She finally made it to Paris and began a job as a seamstress, something she was clearly very good at because it wasn't long before she took part in a royal wedding.

Jeanne was entrusted with embroidering the wedding dress of an Italian princess, Marie-Caroline, the Duchess of Berry, for her marriage to Charles Ferdinand D'Artois, a nephew of King Louis XVIII.

Among the congregation of French and Italian dignitaries was James Power, originally from the British Caribbean colony of Dominica, who had become a wealthy merchant based in Sicily. In 1818, two years after meeting at the royal wedding, Jeanne and James were married in Sicily. The couple settled in Messina, a port on the east coast, where Jeanne became a lady of leisure. She no longer sewed or embroidered dresses for a living, and she didn't continue with such genteel pursuits to keep herself busy, as most other aristocratic ladies were expected to do. Instead she rolled up her sleeves and became a scientist.

On Jeanne's doorstep was the Strait of Messina, the narrow gap between Sicily and the Italian mainland that connects the Ionian and Tyrrhenian seas. For mariners this is a dangerous place where ferocious currents race north and south, switching direction every six hours and sucking tides swiftly up and down. Much feared since classical times, the strait's raging whirlpools and rocky reefs were personified as two sea monsters in Greek mythology, Scylla and Charybdis.

The six-headed shark-toothed beast, Scylla, guards one side of the strait. In Homer's epic poem *The Odyssey*, the hero Odysseus narrowly escapes being devoured by Scylla, although several of his companions aren't so lucky. In a later encounter, Odysseus drifts back through the strait on a raft and this time gets a bit too close to Charybdis, the whirlpool, who sucks up masses of water into her enormous mouth along with the unfortunate Odysseus; he clings on to his raft and waits until Charybdis belches him back out, spinning whirlpools across the sea. In another ancient story, Jason and his crew of Argonauts sail through the perilous waters between Scylla and Charybdis on their way back from stealing the Golden Fleece. They only survive their encounter because Jason convinces Thetis, a sea nymph, to guide the way.

When Jeanne arrived in Messina and began pondering the legendary strait, she didn't go hunting for menacing beasts such as Scylla or Charybdis. Instead she became entranced by some of the real creatures that inhabit these turbulent waters.

Jeanne's interests in the natural world had already begun to lead her around the island, which she would explore for the next 20 years. She wrote a guidebook to the island's wildlife, and studied caterpillars and butterflies, starfish, crabs and even Noble Pen Shells; she described watching an octopus wedging a stone between the pinna's twinned shells before devouring the mollusc inside. Way ahead of her time, she came up with the idea of restocking overfished rivers with fish and crayfish. She also tamed a pair of pine martens that lived in her house while she observed their behaviour; she brought a tree inside for them to climb in, and live birds and squirrels for them to hunt. And it was a curious marine species that tempted her to embark on a revolutionary study. Jeanne realised that she was in the perfect place to answer a time-worn question: do argonauts borrow, steal or make their shells? She knew that to find an answer she had to do something no one else was doing. She would spend a lot of time with living argonauts.

Jeanne had a ready supply of these animals from the seas around Sicily; fishermen sometimes snagged argonauts in their nets and gave them to her, and she also ventured out and caught them herself. All she needed was a way of keeping them alive while she observed and experimented with them, so she devised a series of brand new observation platforms.

One was a simple box, later dubbed the 'power cage': four metres (thirteen feet) wide, two metres (six feet) tall and a metre (three feet) deep, with a door that flipped open on top and two glass observation windows so she could peer in. At each corner was a small anchor that fixed the contraption to the seabed at the shoreline. The cage walls were made of narrowly spaced bars that kept a fresh supply of seawater flowing through, but held the argonauts and their shells captive inside. Jeanne also built a glass tank in her house. It was the world's first aquarium. Her inventions let Jeanne observe the marine world in a way that no one had ever done before, and she settled in for months and years of patient observation and learning.

Watching adult argonauts swim around her aquariums, she saw how easily they climbed all the way out of their shells, and how they aren't permanently fixed inside like all the other molluscs with shells, including the chambered nautilus. She saw how the argonauts held on to the shells with their suckered arms, and noted that they never abandoned their shells altogether.

This was one of the pieces of accurate biology that Jules Verne included in *Twenty Thousand Leagues Under the Sea*. Professor Aronnax tells his servant, Conseil, about argonauts never choosing to leave their shells, even though they're free to go at any time. In reply, Conseil remarks that Captain Nemo should have called his ship not the *Nautilus* but the *Argonaut*, because he too could leave, but chooses to stay confined inside.

Argonauts were clearly different from the other cephalopods Jeanne studied. Placing common octopuses inside her aquarium, they swiftly munched any food on offer before slipping their soft, unshelled bodies through the bars and slinking off into the open sea. This was something the argonauts never did, choosing to hang on to their shells and remain stuck behind bars. And when Jeanne took away their papery spirals, the argonauts died. She concluded that if they were borrowing shells from other animals, then surely they would have wandered outside the cage to try to find another one.

Fracturing their shells and leaving them in place, Jeanne saw that even though argonauts can't make new shells, they do know how to a mend a broken one. Her injured argonauts rubbed the surface of their shells with silvery, web-like membranes at the end of two of their arms, which exuded a sticky substance that sealed up the cracks. Analysing the glue's chemical make-up, Jeanne matched it to the calcium carbonate of the original shell.

Next, she tried breaking off small chunks of shell. After being inflicted with this new level of damage, an argonaut would spend hours sorting through bits and pieces on the aquarium floor, testing out shell fragments to find ones that were a perfect fit for the gaps; it would then glue the chosen pieces in place on its shell to complete the broken jigsaw puzzle.

The discovery that argonauts are equipped with these expert shell-fixing skills lent more support to Jeanne's argument that they do indeed make their own shells and don't simply steal them from other animals. But there was still one final part of the picture left to find: Jeanne needed to catch an argonaut in the act of actually making a shell.

Contrary to the reports made by Giuseppe Saverio Poli, Jeanne saw no sign of a shell when she examined unhatched argonaut eggs. However, she carefully watched as they hatched and grew up, and saw that when the young animals reached the size of a little fingernail, around nine millimetres (a third of an inch) across, they began to build their hard outer covering. As the argonauts got bigger, so did their shells.

Thanks to Jeanne's extensive research, there was no longer any doubt that argonauts do indeed make their own shells, and that they do it in a completely different way from all the other molluscs. Instead of secreting a shell with their mantle, argonauts have shell-making glands at the end of those two arms that she observed repairing breakages; these are spread out into broad membranes (the very same 'sails' that Aristotle imagined argonauts unfurled to propel themselves over the seas).

All of these discoveries could have been lost and forgotten had Jeanne not kept up with her correspondence, and been good at publishing her findings. When Jeanne and James Power decided to leave Sicily and live in London, then Paris, they travelled overland. Meanwhile Jeanne arranged for the bulk of her papers and research equipment to be sent on afterwards by sea; everything was packed up and loaded onto a sailing ship bound for London. A short way into the voyage, off the French coast, disaster struck. The ship sailed into a storm and sank, sending Jeanne's treasured collections into the ocean depths (the kind of romanticised disaster that rarely strikes scientists today, but perhaps a reminder to do regular data backups – the modern equivalent of avoiding a shipwreck).

Jeanne's findings live on in pages of letters she wrote to Richard Owen at London's Natural History Museum, and in the various papers and studies presented in scientific journals. However, her legacy as an early female scientist has faded and she is little remembered for her achievements. Her dedicated research saw her elected as a rare female member of many scientific institutions in Italy, France, Belgium and England, including a corresponding member of the Zoological Society of London. Few people nowadays have heard of the lady who spent years watching, probing and asking questions about this obscure but captivating group of animals.

In the years since Jeanne Power conducted her studies in Sicily, knowledge of argonauts and their way of life has continued to grow. We know they feed on various other animals that live up in the water column, including fish, jellyfish and sea butterflies; we know the females make their shells by laying down material on the inside and outside; we know that no two argonaut shells are exactly the same because of the way they patch them up (this makes it extremely difficult to identify species based on their shells alone); we know that when argonauts meet they sometimes cling to each other and form rafts. No one really knows why they do this, but it could explain why hundreds of them sometimes strand together on beaches.

We also know a lot more about the argonauts' strange sex lives. Throughout her studies, Jeanne noted that she only ever found egg-producing female argonauts. Where were the sperm-making males? She was the first scientist to suggest that the worm-like objects found stuck to female argonauts could be something to do with the males. When Georges Cuvier originally spotted this peculiar appendage he identified it as a parasitic worm, and in 1829 named it *Hectocotylus*. Much later, following Jeanne's suspicions, it transpired that these were not in fact worms at all but important mementos left behind by inconspicuous males.

Without doubt the less impressive of the sexes, male argonauts can be 12 times smaller and weigh 600 times less than females; they barely reach the size of a peanut. The males don't make shells, but they do have an impressive trick up their sleeves. One of their eight arms is specially modified into a sperm-delivery organ. In other words, they have a penis on the end of an arm. What's more, the male argonaut's penis is detachable.

The word *hectocotylus* is now used for the arms of many male octopuses and squid that dole out packets of sperm to females. Amid a grabby clinch of arms and tentacles, the male will reach into the female's body (in argonauts there is a cavity under the mantle; in other octopuses the male pokes into the female's body just under her eyes). He detaches his wriggling, sperm-laden limb, which clamps on to her with suckers. Female argonauts will often collect and carry around the offerings from several males at once.

After dropping their penis some male cephalopods will grow a new one, but not male argonauts. They only get one shot. Their arm drops off, hopefully stuck to a receptive female, and shortly afterwards they die. Female argonauts, on the other hand, keep going and unlike their octopus cousins they can rear many clutches of young over the course of their lifetimes. Most mother octopuses deposit their eggs inside caves and crevices on the seabed. They will usually stick around to watch over their offspring, to fend off predators and keep their broods well oxygenated with wafts of clean water. A deep sea octopus has recently been seen in the Monterey Canyon off the Californian coast guarding her eggs for 53 months, by far the longest period of egg-brooding ever seen in any animal. After all that time, and possibly even longer, she will most probably die, as most female octopuses do, after their single, tremendous reproductive effort.

Living up in open water, where there are no caves to lay their eggs, female argonauts make their own portable, protective nooks to nurture their young. But their shells aren't just brood chambers, they do other things besides. Watching argonauts for brief periods in aquariums, some scientists have argued that air trapped inside their shells is nothing but a nuisance, making it difficult to steer and stranding the animals at the water surface. Others have entertained the possibility that argonauts wilfully manipulate air bubbles inside their shells, and use them like underwater blimps.

It wasn't until 2010 that this matter was put to rest, when Julian Finn from Museum Victoria in Melbourne, Australia paid a visit to the Sea of Japan. Three female argonauts were caught in fishing nets offshore and brought into Okidomari Harbour, where Julian climbed into his scuba gear and carefully took the argonauts with him down beneath the waves. He emptied all the air out of their shells and released them, one at a time, and watched while all of the argonauts performed the exact same routine.

First the argonauts zipped straight upwards, squirting themselves along using jet propulsion. Arriving at the surface, they squeezed out an especially vigorous jet of water that let them bob up and draw as much air into their shells as possible. Next, the argonauts repositioned their funnels and jetted back down, pushing themselves deeper and deeper.

Being essentially open to the water and not fully sealed off, the air bubbles inside their shells were squashed, and shrank as the argonauts swam down and the pressure around them increased. Eventually the argonauts reached a depth where the air volume inside their shells cancelled its weight and the animals became neutrally buoyant, and therefore effectively weightless: they didn't sink or float but hovered in the water column. On reaching that magic depth, between seven and eight metres (about 25 feet) down, the argonauts scooted off horizontally at high speed, swiftly outswimming Julian and his diving assistants.

Watching them disappear from sight, Julian was certain that the argonauts were deliberately using air as a tool to help them swim efficiently at a shallow depth beneath the sea surface, where they are less likely to get knocked around by waves or picked off by a hungry seabird from above. It would explain why the exhausted argonaut brought to the Cabrillo Marine Aquarium needed a helping hand to fill up her shell at the surface and gain some much-needed buoyancy.

Modern genetic studies confirm that the paper and chambered nautiluses are only distant cousins. Argonauts are without doubt octopuses, members of the coleoid lineage alongside cuttlefish and squid. And the nautiluses are the last few survivors of an ancient cephalopod pedigree, the nautilids, that have been doing their own thing for more than 400 million years.

After all that time, these two groups of animals are living proof that having a gas-filled shell is an efficient way of moving through the oceans. They may not be as agile and swift as some of their cephalopod relations, but they are certainly not as primitive or outdated as the label 'living fossil' implies. And we now know for sure that when nautiluses die and leave their shells behind, argonauts don't pick them up and use them.

But humans do.

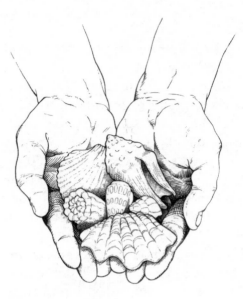

CHAPTER EIGHT

Hunting for Treasures

In the silver gallery at the Victoria and Albert Museum in London – among the hundreds of gleaming goblets, crowns, platters and spoons – is a nautilus cup. It is made from the shell of a chambered nautilus that lived around 400 years ago, and since its death has been transformed.

Most of the shell's ochre stripes have been scraped carefully away to reveal the gleaming mother-of-pearl underneath; glimpses of tiger markings have been left here and there, woven into a swirling design carved across its surface. On the shiny parts a gathering of animals are engraved in fine detail: spiders, wasps, moths and ladybirds. The shell is cradled in a silver gilt mount decorated in enamelled flowers and tendrils, with more insects clambering through them.

The nautilus cup was made in the Netherlands in around 1620. It was probably never actually used as a vessel but would have been put on display in a cabinet of curiosities. Chambered nautiluses were considered to be masterpieces of nature, but they were still something that man could improve on. Just as Noble Pen Shells were displayed alongside articles woven from sea-silk, complete nautilus shells were arranged next to crafted nautilus cups, encouraging viewers to contemplate nature's raw materials and the skill of the artisan who enhanced the shell's beauty through carving and engraving.

The museum's collection includes several more nautilus cups. The Burghley Nef is a French sixteenth-century salt cellar, crafted from a nautilus shell into a medieval sailing ship propped up on a silver mermaid; there is a sixteenth-century cup from England with a golden sea monster opening its ferocious jaws and poised to engulf the tiny figure of Jonah (the original nautilus shell was lost and is now replaced with a silver facsimile); a Polish nautilus shell clasped in an extravagant gold mount and covered with engraved glass and gemstones is dated to 1770.

All of these nautilus shells were probably imported from Indonesia by Dutch merchants aboard some of the same fleets that carried billions of Money Cowries from the Maldives to exchange for slaves in West Africa. These global trade routes supplied an enormous demand for exotic objects from faraway places, including many varieties of seashell.

In auction houses across Europe throughout the eighteenth century, shell mania took hold as rich collectors paid exorbitant prices for rare and beautiful specimens. By the beginning of the nineteenth century, cabinets of curiosities were gradually being replaced by more orderly, systematic collections of natural history objects kept by people who knew what they were looking at. While most collectors didn't stray further than the auction rooms, there were those who aspired to go on much greater adventures. At around the time that Jeanne Power was embarking on her studies of argonauts and their shells, another unsung pioneer of natural history was setting out to pursue an eccentric dream. Hugh Cuming spent years on a series of intrepid adventures on the high seas, risking his life in far-flung lands, and all because he wanted to collect more seashells than anyone ever had before. He brought thousands of shells back from his global journeys; they redefined the boundaries of species diversity in the natural world.

One of the strongest connections many people feel to seashells is the urge, now and then, to collect them, and it's a time-honoured hobby. One of the oldest known shell collections was preserved in the Roman city of Pompeii. When Mount Vesuvius erupted in 79 AD, it choked and buried the city and its inhabitants in ash. Inside one excavated house, archaeologists found a gathering of shells that came from distant seas, certainly as far as the Red Sea, and they seem to have been kept for the simple reason that they looked pretty.

Anybody who has visited a beach has probably spent time idly browsing the shoreline, poking through flotsam and jetsam, to see what the sea has pitched up. Beautiful, spiralling shells are no doubt among the greatest of beachside treasures. They appeal to the hoarder in us all, the part of us that wants to have and keep things, especially those mementos that remind us of a different place and time, of holidays and sea breeze and sand between our toes.

Then there are people who take shell-collecting much more seriously, the ones who get infected with the need to hunt down new things, to write lists and keep scores. The thrill of discovery was probably what drove Hugh Cuming to do what he did.

As a young boy, he explored the beaches of south Devon, on the heel of England's south-westerly foot that points towards France. He was born five miles inland, at the end of a winding estuary in a hamlet called Dodbrooke, on St Valentine's Day, 1791. His siblings were Jane, Thomas and James, Richard and Mary were his parents, and little else is known about Cuming's early home life. He was only one when his father died, and by his teens he was apprenticed to a local sailmaker, where he learnt a profession that would eventually lead him to the other side of the world. For now, though, Cuming stayed close to home, where he may well have encountered a trio of men who lived nearby: a magistrate, a colonel and a shoemaker. All three were adventurers in their own way, and together could have shown the young Cuming what possibilities the world had to offer if you just went to look for them.

The first of these men was Charles Prideaux, a gentleman who lived most of his life in a fine, stone-fronted house smothered in vines in the centre of Kingsbridge, a town near Dodbrooke. Prideaux was a magistrate, but his heart lay in the natural world, as it did for many others of his generation. He belonged to a clique of amateur naturalists that swelled in number greatly during the eighteenth century. For those with a little spare time there was no better hobby than gathering fascinating and beautiful objects from the natural world. Prideaux was especially enchanted by animals with shells. He made grand collections of seashells and crabs, and developed a special fondness for bizarre creatures that combine the two – the hermit crabs – including several new species that were named after him. It's not known whether Cuming ever met Prideaux, but he may have heard stories of the ardent naturalist rowing out into Plymouth Sound and lowering a small wooden dredge into the depths to bring back hidden wonders from the seabed.

Cuming seems to have made a personal connection with another local naturalist. Colonel George Montagu retired from a long army career to live in Kingsbridge, where he wrote books about birds and molluscs, including hundreds of species that he spotted in Britain for the first time. According to several accounts, Montagu took the young Cuming under his wing, encouraging him to explore the Devon coast and start his first shell collection.

There was one other Kingsbridge man who may have inspired Cuming to pursue more exotic adventures. Compared with the other, wealthier naturalists, Cuming had more in common with John Cranch. Both Cranch and Cuming were sent at a young age to learn a trade – Cuming made sails and Cranch shoes – and both of them turned out to have an adventurous spark.

John Cranch was desperate to become a full-time naturalist. He worked hard at his Kingsbridge shoe business to make ends meet, but escaped whenever he could to sea. He assisted Colonel Montagu, often accompanying him on dredging trips offshore. He wrote articles and papers about his findings and discovered new species that were named in his honour.

In 1816, after his friend William Leach at London's British Museum put in a good word for him, Cranch was taken on as the zoologist on a Royal Navy expedition to find the source of the River Congo. The voyage got off to a bad start when the brand new, 30-tonne, 20-horsepower steam engine only propelled the vessel at three knots, barely faster than strolling pace. The paddle wheel and engines were stripped out, and HMS *Congo* finally cast off under sail power. On the way to Africa, Cranch gathered zoological specimens, including the living argonauts that would later bear his name. When they eventually reached what is now the coast of the Democratic Republic of Congo, the ship only made it a few hundred miles inland before impassable waterfalls and rapids blocked their way. The only thing the crew discovered about the origins of the Congo River was that the only way to find it would involve a lot of walking. They struck out overland on foot but a terrible illness soon broke out, probably yellow fever. Cranch fell ill, and for 10 days he was slung in a hammock and carried back to the ship where he soon died, along with more than half the crew.

Shortly before grim news of the Congo expedition filtered back to England, Colonel Montagu met a far less exotic but equally fatal demise. He stepped on a rusty nail at his house in Kingsbridge and died of tetanus; staying at home or exploring faraway lands, either way life was precarious at the start of the nineteenth century. It is easy to imagine which of the two fates Hugh Cuming would have wished on himself, if he had had to take his pick, because he soon set off on overseas adventures of his own. In 1819, aged 28, he left Devon for the first time and set sail for the southern hemisphere to take a job as a sailmaker in Valparaiso, a major seaport on Chile's mid-west coast, where a new British colony was growing fast.

Life went well for Cuming in Chile. He met Maria de los Santos, who he never married but their daughter, Clara Valentina, was named in honour of her father's birthday. In his spare time, Cuming scoured the rocky shores and inlets around Valparaiso, and began to amass a considerable collection of shells that were new to him. Both his work and his hobby introduced him to various local characters – port inspectors, customs officers, bureaucrats and fellow shell collectors – who would prove to be immensely helpful in the years to come. It was one of them, a Lieutenant John Frembly, who announced Hugh Cuming to the scientific world in 1825 when he described a new species of chiton.

'I have named this species after my friend Mr Cumings,' wrote Lieutenant Frembly. He went on to speculate that Cuming would 'soon make a large addition to our present stock'. He may have misspelled his friend's name but Frembly was not wrong in his hunch that there was much more to come from this enthusiastic collector. It would take a few years for the world to find out just how substantial Cuming's contribution to shell-collecting and science would be.

Cuming had done very well as a businessman in the short time he had lived in South America, and by 1826, aged only 35, he had built up enough savings to retire and devote himself to chasing a grand ambition. Cuming built a small wooden schooner, decking it out with collecting kit and ample storage space. It was probably the world's first custom-made vessel devoted to scientific research. He hired the services of a Captain Grimwood, and on 28 October 1827 the two of them cast off the ropes of the *Discoverer*, waved goodbye to Maria and Clara, and set sail due west to see what they could find – including as many shells as possible.

By the time Cuming and Grimwood sailed into the Pacific, a new age of scientific discoveries was well underway. Up until the turn of the eighteenth century, explorers travelled the world mainly to try to acquire and expand colonies and to open up new trade routes. Political and economic ambitions never went away, but they were joined by a growing scientific curiosity guided by a new fellowship of scientists. Professional societies were forming in cities across Europe, and they were the driving force behind many great expeditions; scientists became indispensable members of the crew.

Captain James Cook was hired by the Royal Society in London in 1768 to sail to the Pacific Ocean to observe the transit of Venus across the face of the sun. On board with him on HMS *Endeavour* were the naturalists Joseph Banks and Daniel Solander, who were in charge of gathering plant and animal specimens along the way, including a lot of seashells. Solander was one of 17 young adventurers recruited by Carl Linnaeus to join expeditions around the world, to collect specimens and test out and expand his new binomial formula for naming species (giving them a two-part name, first genus then species, as in *Homo sapiens*). Collecting animals and plants and cataloguing them according to

Linnaeus's new scheme became a major goal of eighteenth-century exploration, and many global voyages returned with hoards of natural history specimens. The French ships *La Boussole* and *L'Astrolabe* set off in the 1780s with the aim of completing Cook's exploration of the Pacific, but they both vanished without trace in the Solomon Islands. Numerous voyages attempted to locate the Northwest Passage that was believed to connect the Atlantic and Pacific by a northerly route; other trips made detailed observations along the coasts of India, China and Australia.

All these globetrotting efforts helped uncover a simple and powerful truth about the natural world's biological riches: they showed that patterns of life vary across the globe. In order to find new varieties of plants and animals, simply go and look carefully in places no other scientists have been before. New places: new species.

The findings of the early scientific expeditions, Cook's voyage in particular, no doubt gave Hugh Cuming the idea of sailing across the Pacific in search of new and unknown shells. What set Cuming and Grimwood apart from other collecting expeditions at the time was the small scale of their mission. It was just the two of them. There was no big ship filled with provisions and a permanent, supporting crew on hand, and no money from government or scientific societies; just Cuming's private funds, and the hope that when he got back he could sell some of his shells while keeping the best specimens for himself.

For eight months, Cuming and Grimwood island-hopped across the Pacific on board *Discoverer*. Cuming chronicled their voyage in a journal, of which a copy survives (he probably wrote it up on his return to Chile). It traces their route, and offers glimpses into the other adventures they had along the way besides shell-collecting.

It took them a week to sail 400 miles to their first stopping-off point, the Juan Fernandez Islands, famous as the home of castaway Alexander Selkirk, the inspiration for Daniel Defoe's *Robinson Crusoe*. Selkirk was rescued 100 years before Cuming and Grimwood called in; unlike Selkirk's three years of isolation, they stayed for just a week. During that time Cuming got his collection underway and he had already found some shell species that were different from those he knew from the Chilean coast. He also noted an abundance of goats left behind by visiting sailors and pirates, and lush vegetation with fruit and vegetables introduced from Chile, including 'radishes of an extraordinary size'.

The *Discoverer* next called in at Easter Island, where Cuming began his collection of anthropological artefacts, bartering cotton handkerchiefs for small wooden idols carved by the locals. He had brought with him a stock of tobacco, wine and colourful ribbons, which he exchanged throughout the voyage for traditional weapons and musical instruments; he was especially fond of his two nasal flutes. Throughout his journal, it is clear that Cuming was fascinated by people and places; he goes into great detail on the local costumes and practices, buildings and food. On Easter Island Cuming found more shells, saw the monumental moai statues, and stocked up on fresh provisions before heading onwards into the Pacific.

By December, they reached Pitcairn Island, calling in on John Adams, the last of the mutineers of HMS *Bounty,* who had sought refuge on this remote volcanic outcrop several decades earlier. After a few uneventful days in Pitcairn, they left behind the remote reaches of the central Pacific and made their way to a string of idyllic, palm-fringed islands and coral atolls that nowadays lure in legions of sun-seeking holidaymakers. Back when Cuming was there, the only visitors to French Polynesia were whaling ships and the occasional naturalist passing through, as well as the Christian missionaries who came and stayed.

Not everywhere they went were Cuming and Grimwood welcomed by friendly locals and expat missionaries. In some places they were met with ferocious war dances, blood-curdling yells and displays of menacing weapons. On Temoe Island, in the south-eastern fringes of the Tuamotu archipelago, a clumsy scene unfolded that could have ended in disaster. Cuming and Grimwood, along with four locals hired from a nearby island, were rowing in a small boat towards the beach when two islanders spotted them and dashed down to the water's edge, spears in hand. It was Captain Grimwood's idea to frighten them off by firing a few musket shots over their heads; all this did was draw a bigger crowd of shouting, dancing men wearing war helmets topped with feathers, their bodies painted black and white.

Spying a gap in the coral reef on the other side of the lagoon, Cuming's small crew tried to make their escape but encountered a wide stretch of dry sand and rocks between them and their exit. Their only option was to pick up the boat and scuttle as fast as they could towards open water while the islanders swarmed after them.

Even when they reached the water Cuming and his men were still not in the clear. A wave broke over the side and capsized the boat, scattering their belongings in the sea. Cuming was flipped out of his seat, the boat landed on his leg, knocked him unconscious and he sank to the bottom.

Rescue came from one of their hired hands, who swam down and dragged the sodden Cuming back to the surface. The crew righted the boat, only for Cuming to get washed overboard, and rescued, a second time. The islanders ignored the commotion in the sea because they were too busy fishing out the hats, jackets, oars, collecting baskets and bottles that had fallen from the rowing boat and were floating towards the beach.

It's quite possible that Cuming and Grimwood were imagining the savagery of the Temoe islanders, and clearly the two groups of people were bemused and confused by what the other was up to. As the visitors limped away using the single emergency oar they found strapped to the boat, two islanders ran along the beach after them waving the jettisoned oars. They threw the oars in the water and Grimwood ordered a strong-swimming member of the crew to go and fetch them. As he did, one of the islanders also jumped in the water, frightening the boat boy into turning tail and scrambling back to the boat as fast as he could. Cuming left Temoe with a single shell that he had found under a stone and that somehow hadn't fallen out of his pocket during his multiple dunkings in the sea.

In February 1828, after finding lots of pearl oysters in the lagoon of South Marutea Island, Cuming halted the expedition for a month, built a small house under the palm trees and hired a team of men to dive for pearls. He writes in his journal about several other Pacific lagoons that had already been stripped of their pearls. Nevertheless, his men gathered 40 tonnes of oysters and 27,000 pearls from South Marutea, although they were later deemed to be too small and ugly to have any great value in Europe.

The *Discoverer* called in at several other islands across the Tuamotu archipelago, all of them sandy atolls ringed by limpid lagoons and fringed by coral reefs – Tureia, Nengonengo, Motutunga, Anaa – with Cuming adding to his collections at each one.

By April, Cuming and Grimwood had arrived in Tahiti, where they were welcomed by Queen Pomare. The 15-year-old royal, along with the queen mother and several attendants, all boarded the *Discoverer*. They were accompanied by Mr Kimpson, a local missionary, and while he was there the royal party behaved themselves, genteelly sipping fine Chilean wines and indulging in a little light conversation. But as soon as Mr Kimpson left, things got considerably more lively. Bottle after bottle of wine was drunk and the guests showed no signs of wanting to leave. After dinner, the tipsy queen collapsed in one of the *Discoverer*'s bunks and slept off the partying, while her escorts waited patiently on deck until sunset, at which point the ladies finally left Cuming in peace. The following day, back on land, the royals had recovered from their hangovers and welcomed Cuming and Grimwood with lavish tropical fruits. They also agreed to the men's request to halve the normal duty on visiting ships from twelve to six dollars, on account of the *Discoverer* being such a small vessel (although roomy enough for a good knees-up).

Tahiti was the westernmost point on Cuming's Pacific journey, and it proved to be a treasure trove of shells; his journal records 98 species that he hadn't already found elsewhere on the trip. Swinging the bows of the *Discoverer* eastwards, they began the 5,000-mile trek back to Chile, stopping off at more islands, and all the while Cuming's collection continued to grow. By the time they arrived back in Valparaiso in June 1828 they had visited more than 50 islands, weathered only a single storm, met hundreds of missionaries and Pacific islanders, and laid the foundations for a shell collection that would transform the world of conchology.

Cuming continued his explorations on a subsequent voyage along the Pacific coast of Central and South America. Much less is known about this trip since no journal survives, but piecing together a picture from letters he wrote, it is clear that he once again ventured on board the *Discoverer* and followed the course of the Humboldt Current, a cool oceanic river that sweeps up the coast from Chiloé Island in southern Chile to Peru. He continued north into Panama and Costa Rica, Nicaragua and Honduras, then picked up the trail of the Humboldt again as it swings offshore towards the Galápagos Islands, where Cuming arrived some two or three years before Charles Darwin on HMS *Beagle*. The paths of these two men would cross again, and more closely, in the years ahead.

One thing Cuming did differently on this trip compared to the eastern Pacific voyage was to use a dredge, presumably an idea he had brought with him from Devon and his mentor Montagu. Rather than just collecting by hand and

occasionally hiring the services of local skin-divers, Cuming
fixed a small dredge behind the *Discoverer* and towed it along
to bring up samples from much deeper down. In a letter
written years later to Edgar Layard, a fellow collector,
Cuming recommended: 'You must carry with you when
you go dredging a fine sieve, a hand bucket, and a large
coconut shell.' The coconut shell was to scoop off the mud
and sand, and the sieve to gently sift out the shells, including
tiny specimens that were the treasures of the dredge-spoil
and otherwise virtually impossible to collect. Cuming also
gave instructions on how to prepare shells for transport; one
should boil bivalves alive then pick the animals out, and
carefully tie their shells closed with a piece of string;
gastropods can be left in a glass jar to rot for a month or so,
somewhere 'the stench will not offend'. The snails entirely
decompose, leaving their shells ready to be washed clean.

Throughout his Latin American travels, Cuming was
assisted by letters of recommendation from Chilean
dignitaries that he met along the way and clearly struck up
good friendships with; he was also granted exemption from
port fees and taxes. However, in Jipijapa in southern Ecuador
he did run into a spot of trouble. Cuming was arrested and
thrown in jail. Local authorities somehow mistook his little
boat for a Peruvian frigate, and didn't want to take any
chances since Peru had recently laid siege to the city of
Guayaquil. The incarcerated collector calmly explained that
his boat was far too small to be a frigate, and his only
scheming was of the molluscan variety. They let him go,
shaking their heads in disbelief at his devotion towards such
apparently insignificant creatures.

Cuming would begin to discover the true value of his
collection on his return to England in 1831. He left Valparaiso
and never went back, even though by then his mistress
Maria had given birth to a son, Hugh Valentine.

In London, Cuming immediately immersed himself in
the gentleman's world of conchology. From the outset, he

resolved not to write anything about his shells himself; he never described or named a single one. He would leave that to the experts. As he saw it, his role was simply that of collector, to provide material for others to work with.

His collection was always open to any and every naturalist who wanted to use it. He mailed shells to experts in America, he welcomed visitors to his home and he set about a lifelong collaboration with several prominent gentlemen in London.

In February 1832, a selection of Cuming's shells was put on display at a meeting of the newly formed Zoological Society of London. The shells were accompanied by drawings and written descriptions by George Brettingham Sowerby and William Broderip (the same man who had written about argonauts and 'fairy boats' a few years previously). They would become two of Cuming's most trusted friends. Broderip and Sowerby (later followed by his son and then his grandson, G. B. Sowerby II and III) sat down to describe, draw and name thousands of Cuming's shells. In 1832 alone, in the pages of the Zoological Society's journal, Broderip named 247 new mollusc species from the great collection, and he continued to add many hundreds more each year.

A few months after those first species were named, Cuming was elected as a fellow of the Linnaean Society of London, a distinguished institution dedicated to the study and discussion of natural history. It was a triumph for this unschooled boy from Devon, but he was far from finished with his conchological adventures. He spent a few years in London, selling his duplicate shells at auction houses and acquiring new specimens of species he hadn't already found himself. Then he gave in to his itching feet, and began making plans for a third great journey.

Into the Coral Triangle
Cuming decided to go to the Philippines, a cluster of islands in the far west Pacific that naturalists were just beginning to explore. And it was a perfect destination because, even though

it wasn't known at the time, this archipelago is crammed with species and many are endemics, found nowhere else on the planet.

The islands lie within a region that has come to be known as the Coral Triangle. A rather misshapen triangle admittedly, it stretches from Papua New Guinea and the Solomon Islands in the east to Bali, Kalimantan and Sabah in the west and northwards to the Philippines. This is *the* global epicentre for marine biodiversity, home to 40 per cent of all fish species in the world, and three-quarters of all the coral species; in the Coral Triangle, one hectare (2½ acres) of reef (the same area as Trafalgar Square in London) contains more coral species than the whole of the Caribbean Sea. That's not to mention six of the world's seven sea turtle species, dozens of marine mammals and a throng of other varieties of life all crammed in together. If an antique map showed where all the sea creatures live, there should be a label pointing to the Coral Triangle saying 'Here be beasts (*lots* of beasts).'

Scientists are still trying to explain this remarkable phenomenon. Theories include the possibility that the Coral Triangle is a 'cauldron' of speciation, where more species have evolved than anywhere else. Equally it could be that fewer species have gone extinct than in other areas. Alternatively, species that evolved elsewhere could have accumulated in the Coral Triangle, either by drifting on ocean currents or by moving with the slow shift of islands on drifting continental plates. A fourth option is that the Coral Triangle is a region of overlap between species in the Indian and Pacific Oceans, like the middle part of a Venn diagram. It's not known if any of these ideas are correct; perhaps there is no single reason but instead a mix of many different things going on. Even with modern techniques, including genetic analyses looking at how species are related, the puzzle of the Coral Triangle and its outrageous biological riches still remains unsolved.

When Cuming and Grimwood were sailing westwards from Chile into the Pacific, they were following a gradient of increasing diversity; the closer they approached the Coral Triangle, the more species they were likely to encounter. Arriving in the Philippines, Cuming had jumped right into the middle of things. Around 3,500 species of marine molluscs are known to live in these islands. Add in the undiscovered species and the estimated total reaches 15,000 in shallow waters, with another 20,000 deeper down, plus many thousands more on land. Cuming was not going to have any trouble finding new and interesting molluscs in the Philippines.

In January 1836, Cuming set sail for Manila, although not on board the *Discoverer*, for this voyage, he would island-hop in relative comfort as a guest of the Spanish government, who at that time ruled the Philippines. Spanish priests around the islands provided him with places to stay, large boats to sail on and hordes of eager schoolchildren, who were conscripted to his collecting efforts.

Few details are known about the three and a half years Cuming spent in the Philippines. He did write a journal but, as we will see, there are no known copies. Letters to friends and scientists back in Europe, and accounts of his life written by those who knew him, give a few vignettes into his time in the region.

As in his previous journeys, local residents in the Philippines were puzzled by what Cuming was up to. They wondered why this gentleman from Europe paid people to find shells for him (most white men took money *away* from them). And why did he sit up late into the night cleaning and sorting the shells? Cuming tried and failed many times to explain the enthusiasm back in his home country for collecting natural history specimens. In the Philippines, Cuming saw people had a rather different use for shells: they burnt and crushed them, mixed them with betel nut, wrapped them in leaves and chewed them (betel nut remains the fourth most widely used drug around the world, especially in Asia, after nicotine, alcohol and caffeine; the burnt shells produce calcium hydroxide, which helps extract the active chemicals in the nut). Eventually Cuming gave in and told people he was planning to sell his shells to Europeans who had the same nut-chewing habits.

It was while Cuming was in the Philippines that he came across a molluscan superstar. The Glory of the Sea, *Conus gloriamaris*, is a fairly large cone snail that grows up to 13 centimetres (5 inches) long and is decorated in immensely fine, golden-brown saw-tooth markings. It is undoubtedly a pretty shell, although no more stunning than hundreds of other cone snail species. What drove collectors to distraction was its rarity. Since its discovery in 1777 only a handful of specimens had been found and no one knew where to find more. The Glory of the Sea became one of the most famous and valuable shells in the world.

In 1824, William Broderip almost paid £99 19s 6d for a single Glory of the Sea, but he was outbid at the last minute by another collector who paid £100 (adjusted for inflation that is equivalent to almost £8,000 today). Possibly apocryphal stories reveal how carried away people got about this particular species. One tells of a Danish collector in 1792 buying a Glory of the Sea at auction and immediately stepping on it, smashing it to bits in front of a gawping crowd simply to make the specimen he already had all the more valuable. Whether the story is true or not, the fact that people told it shows how obsessed they were about the Glory of the Sea.

Another story often told revolves around what happened after Cuming found this rare species in the Philippines. He was collecting along the shore off Bohol Island when he turned over a rock and saw nestled underneath it not one but two, or possibly even three, Glories of the Sea. The story goes that Cuming was so overcome with joy at the sight of these rare shells that he danced about in sheer delight.

On returning to the same spot some time later, Cuming supposedly discovered that the island had been struck by an earthquake, and his valuable collecting spot had sunk down out of reach beneath the waves. At the time, this was the only place in the world where Glories of the Sea were known to live, and now it was lost. The Philippines is well known for catastrophic earthquakes, but we simply can't be sure whether Cuming's collecting spot really was obliterated; perhaps this was yet another story told to cast shadows around these mysterious and valuable shells.

By 1839, Cuming had packed up his collections and was ready for the six-month sea voyage back to London. In a letter to Richard Owen, he reports having more than 3,000 mollusc species; 500 of them were from forests and rivers, the rest from the rich coastal waters. Cuming may have been the first person to export crates of shells from the Philippines, but he certainly wasn't the last. Today, the Philippines is a major hub in a multi-million-dollar global shell trade. This really got going in the 1970s, when the Philippines government encouraged people to gather and sell seashells to the growing number of foreign tourists visiting the islands.

Thousands of shell species are involved in various aspects of the modern shell trade. Some are sold in bulk as the raw material to make mother-of-pearl inlays and jewellery; others, especially cowrie shells, are sold for use in shellcrafts, to make into necklaces, chandeliers and ghastly shell figurines and ornaments. There is also a specialist trade in fine and rare specimens that are treated as gems and sold on specialist networks to discerning collectors around the world. But rather than having to wait for an explorer like Cuming to return from an epic voyage, as European collectors did in the nineteenth century, shell enthusiasts today can simply browse through a website, pick the shells they want and have them delivered from the Philippines directly to their front door.

There was one notable omission among the species and specimens Cuming collected. In his letter to Owen, Cuming

confessed that he hadn't been able to find a whole chambered
nautilus with shell and body intact. Owen had a special
interest in these animals. In 1832, he had published his first
monograph on their anatomy, called *Memoir on the Pearly
Nautilus*. It was based on a single preserved specimen given
to him the previous year by the British naturalist, George
Bennett, who had spent years exploring the Pacific. All
Cuming had managed to find was a few empty nautilus
shells. He could quite easily have brought back another
specimen of this elusive animal from the Philippines for
Owen, if he had only known how to catch one.

A nautilus interlude

To catch a nautilus is fairly straightforward: make a wooden
or wire trap with an entrance just big enough for a nautilus
to swim through; bait the trap with cat food or scraps of
chicken, then lower it down into deep water to at least 100
metres (the method does, admittedly, require a lot of rope)
and leave it overnight. Scavenging nautiluses will pick up
the whiff of food and come to investigate. Once inside the
trap, the short-sighted cephalopods can't find their way back
out. All the nautilus hunter need do is return the following
morning and pull all that rope back up.

This is how the commercial trade in nautilus shells is
carried out today, with tens of thousands of animals caught
and killed each year. Gilded nautilus cups may have fallen
out of fashion, but these days people use the spiralling, tiger-
striped shells to make all manner of other ornaments,
lampshades and buttons. There are even people who eat
chambered nautiluses; between 2007 and 2010, Indonesia
exported 25,000 nautiluses to supply meat markets in China.

The Philippines is a major player in the nautilus trade.
Exports have fluctuated; in 2008 around 54,000 nautiluses
were recorded in trade; the following year the number tripled;
then in 2010 the trade dropped again to around 24,500
animals (figures are grouped together for all the *Nautilus* and

Allonautilus species). A lot of these nautiluses end up in the United States. Between 2006 and 2010 more than half a million items were imported to America, most of them whole shells.

The Philippines was also one of the first places where nautilus fisheries began to collapse. Anecdotal reports suggest that nautiluses have gone locally extinct in the Tañon Strait between the islands of Negros and Cebu. This happened in the 1980s and since then the fishery hasn't restarted, suggesting that the nautiluses have not recovered. If they had, fishermen would surely be going out to catch them.

A 2014 study used video cameras baited with chicken to survey nautilus populations in a fishing ground of the Philippines and compared it to three un-fished sites in other countries. Individual nautiluses were identified based on unique patterns of stripes on their shells. The highest numbers were counted in Australia, at Osprey Reef in the Coral Sea (68 nautiluses) and on the Great Barrier Reef (92). Twenty nautiluses were identified at each of two other un-fished sites, Beqa Passage in Fiji and Taena Bank in American Samoa. By contrast, in the Philippines, the drop-down cameras spotted only six nautiluses. Even taking the soak-time of each camera into account, estimates of the total population are still far lower in the Philippines than elsewhere. Several factors could explain these differences, including variety in habitat type and limitations of the filming technique. But most likely is that the Philippines population has been depleted by fishing while the others have so far been left alone.

These results are no great surprise, given a few basic facts of nautilus biology: they don't become sexually mature until they are teenagers, at least 15 years old; when they do, a female spends a year incubating eggs inside her shell before a meagre 10 or 15 hatchlings emerge. Compared to many other molluscs, there really is no possibility that the ocean will become overrun by nautiluses any time soon – quite possibly the opposite.

Based on this latest research, it seems nautiluses could be far less abundant than anyone ever imagined, even in areas where they aren't being hunted. People have been admiring and collecting their shells for hundreds of years but now the pressure on them could be too high. Many experts are calling for the immediate control of the global nautilus trade to make sure that this narrow branch of the tree of life doesn't face one mass extinction too many, and finally get snapped off.

Of gentlemen and disappointment

Cuming returned to London in June 1840, whereupon he hung up his explorer's hat once and for all. He spent the rest of his life expanding his shell collections from the comfort of his new home at 80 Gower Street in Bloomsbury, a short stroll from the British Museum. He would still visit auction houses and museums across Europe, but never again departed for exotic, faraway shores.

He distributed much of his Philippines material among naturalists and collectors – not just shells but thousands of birds, insects, crabs and reptiles, and 130,000 dried plants, which he hoped would impress William Jackson Hooker, the man who would soon be appointed as director of the Royal Botanic Gardens at Kew, and who Cuming had been writing to for years.

As well as all his plants, Cuming also sent Hooker the journal he wrote during his Philippines sojourn. In the accompanying note, dated May 1841, Cuming refers to his journal as his 'child', apologising effusively for his bad spelling and grammar and asking for help in editing and publishing the work. Unfortunately, Cuming had picked the wrong man to ask for assistance. While Cuming was well liked by many throughout his life, there were a few scientists who didn't seem to take him seriously; Hooker was one of them.

Hooker rejected Cuming's work and the journal was lost, perhaps carelessly by Hooker or deliberately by a disappointed

Cuming. With the journal gone, Cuming's hopes of being fully accepted as a gentleman of science were shattered. Perhaps it was his lack of schooling or his refusal to describe any of his collections himself, but throughout his life and for decades afterwards, there were some shell experts who didn't value Cuming's work. In 1909, conchologist Charles Hedley described Cuming as 'an illiterate sailor' and complained 'his plans did not regard the advancement of science'. Another shell scientist in 1939 described Cuming's collection as a 'pestilential conchological swamp'.

Many argued that Cuming had done a bad job of labelling his collections with localities of where the shells were found, a vital piece of scientific information. These allegations were rather unfair given the nineteenth-century custom of attaching only brief notes to specimens, with locations often as vague as 'South China Sea' or 'India'.

On the contrary, it seems Cuming had an encyclopaedic memory for where his shells came from; the only problem was that he kept most of that information in his head, and not written down. People who watched him at work reported how he would lay out parts of his collection on long tables, then dictate notes to an assistant, all from memory.

There is no doubt that Cuming was incredibly generous with his shells. Dozens of scientists passed through his doors to examine and describe them. One of them was Charles Darwin, and over the course of several years the two men corresponded at length and occasionally met. Cuming identified all of the shells Darwin brought back from the Galápagos, he discussed ideas with him on coral reef formation, and lent him many specimens. One of the most important, although not a mollusc, was *Ibla cumingi*, a barnacle Cuming brought back from the Philippines. In the introduction to his book *A Monograph on the Sub-class Cirripedia,* Darwin thanked Cuming for persuading him to spend time looking at barnacles, and said Cuming had 'placed his whole magnificent collection at my disposal'.

Cuming even let Darwin chop up some of his precious specimens. When he dissected *Ibla cumingi*, Darwin found bizarre miniature male barnacles clamped tightly to the giant females (a little like male argonauts), giving him vital clues as to how sex evolved.

But Darwin wasn't always so admiring of Cuming. In 1845 he described him as 'very difficult to make stick to his work' in a letter to his friend Charles Lyell. By then, however, Cuming's health was failing, perhaps from all the years of tropical exploration, which might explain why he was not paying much attention to Darwin. A short time later, Cuming suffered a stroke from which he was not expected to recover.

In December 1846, a letter arrived at the British Museum from the ailing Cuming offering his great collection of 52,789 shell specimens, including at least 18,867 species. The price for his entire collection was £6,000, equivalent to at least half a million pounds today.

His offer to the museum was followed by letters from several eminent zoologists including Richard Owen and William Broderip, urging the trustees to buy Cuming's shells. They pointed out how bothersome it would be if the collection were broken up and lost overseas, scattering this rich source of study far from British soil. John Edward Gray, keeper of zoology at the British Museum at the time, was less enthusiastic, and perhaps under his influence the museum rejected Cuming's offer.

Despite his illness, Cuming lived on a further 20 years until he was 74, with his daughter, Clara Valentina, now by his side. He continued to add to his collection, funding younger men to go on collecting expeditions for him, and he still paid visits to local auction houses. In April 1865 Cuming was spotted in Covent Garden bidding on shells, and was remembered by one collector as a 'somewhat stout, rubicund, good-humoured looking old man'. A few months later, on 10 August, Hugh Cuming passed away at his home in Gower Street. His hair was a jumble of white curls, his skin creased by years of sun and sea, and he was surrounded by his beloved shells.

By then, his collection included some 83,000 shells, proof that he had surely achieved his lifelong ambition. This was without doubt the largest and most famous collection of shells then in existence. A great number of them were ones he had found himself throughout his extraordinary adventures, exploring places no other shell-collectors had been, island-hopping, dredging the seabed, dipping in rivers, shaking tree trunks and picking over leaves and rocks. While other collectors and museums would eventually bring together more shells, Cuming's is still the most impressive collection that one person has amassed. But he didn't live to witness his final wish coming true, to see his fine collection on display at the British Museum.

In a room lined with tall mahogany cabinets I can't decide where to start. Jon Ablett, one of the curators at the mollusc section at London's Natural History Museum, helps me out and picks a cabinet. He swings open the doors, revealing two rows of drawers with brass name-plate holders on each one. Carefully, I slide out a drawer and find it stuffed full of shells, sealed inside small clear plastic bags and nestled inside open card trays that look like giant matchboxes. Jon rummages through the boxes, pulls one out and puts it on a table top in front of me.

There are two spiralling snail shells, a few centimetres tall, cream-coloured with a brown stripe coiled around them. With them are a few bits of paper covered in minuscule, neat handwriting. Jon explains that no notes are ever thrown away; even when experts re-identify a specimen as a different species they simply add their notes to the paper trail of ideas and discoveries.

A tiny square of yellowing paper falls out onto the table with the letters MC written in fading ink. 'That's how we

know this was one of Cuming's,' Jon tells me. The MC
stands for Museum Cuming, the name he gave to his gigantic
shell collection.

After he died, the British Museum eventually agreed to
buy Cuming's shells for the same £6,000 he originally asked
for. A story has often been told of the day when his shells
were eventually brought to the museum. The weather was
blustery, so the story goes, when John Gray's wife carried
tray after tray of shells across a courtyard. As she went, the air
around her was filled not only with a swirl of autumnal
leaves but with hundreds of paper labels from the shells,
mixing them up and whisking away the names, collecting
locations and all scientific value from his shells. However,
investigations by Peter Dance for his book *A History of Shell
Collecting* revealed these tales to be completely made up. It
wasn't Mrs Gray who fetched the shells, and the labels didn't
get mixed up. Perhaps people spread these rumours to try to
fuel antipathy towards Cuming and his shells.

Back in the museum, Jon and I open up more cabinets
and drawers, and we keep finding more MC shells. Some of
them have reference numbers that identify each individual
shell in the museum's enormous catalogue; in the past this
was a handwritten ledger, which the curators are now
working their way through and digitising. Normally, when
new specimens arrive at the museum they are logged in
with a reference number, but that didn't happen with
Cuming's shells because there were simply too many of
them. Instead they were distributed, species by species,
through the rest of the museum's mollusc collection. Only
when someone takes out one of his shells, studies it, then
writes and publishes something about it is it given a number.
There are still masses of unnumbered shells that haven't yet
made it into print.

'People are still identifying new species from Cuming's
shells,' says Jon. Of those that have been identified, many
bear his name, including Cuming's Cowrie, Cuming's

Scallop and Cuming's *Spondylus*. He has all sorts of other animals also named after him, including a starfish, a gecko, a beetle and a tree-climbing rodent from the Philippines called a cloud rat.

I must admit that I had expected Cuming's shells to be all kept together in one place, but actually I prefer that they've been split up and absorbed into this living, working collection. There are approximately nine million shells in the Natural History Museum – this is one of the biggest mollusc collections in the world – but there's no hiding the fact that nobody really knows how many shells they have. Thousands more are added each year, and Jon and his colleagues have to keep finding more space to squeeze them all in.

The main use of the museum's shell collection is to study the diversity and evolutionary relationships of the immense mollusc lineage. Together these specimens form reference points in time and space that people can keep coming back to in the future, to ask questions no one has thought of asking yet and to answer them in ways that haven't been invented.

It's easy to think of museums as dusty places, frozen in time, but they are constantly changing and embracing new techniques and technologies. Most of the Natural History Museum's molluscs are empty shells, because that was the easiest way of collecting them, and in the past there was no way of extracting genetic information from them. But recent advances in DNA amplification mean that even minute scraps of dried mollusc meat stuck inside a shell can now be used to sequence the animal's genome.

Other unexpected and powerful insights come from whole specimens kept in alcohol. Jon tells me he recently had a visit from researcher Justin Gerlach who came to study the preserved remains of a snail species from Tahiti that went extinct in the wild decades ago. A few living *Partula* snails were taken into captivity for a breeding programme that it was hoped would save them from total extinction, but it's not going well; there are only 15 snails of this particular species left, and even those are now dying. By dissecting the historic specimens from the wild it's hoped researchers can identify their final meal and work out if the captive-breeding efforts are failing because the snails are being fed the wrong food.

It's not just scientists who use the collection. The mollusc department welcomes in all sorts of people: artists, designers and engineers all come through to the back rooms of the museum to learn about and seek inspiration from the shape and form, architecture and beauty of these many millions of spirals.

Art historians also visit the department to examine an extraordinary series of shell books. Lined up along one shelf in the mollusc section's library are 20 huge volumes, their titles and contents embossed in gold on the spines: 'Vol. 1. *Conus, Pleurotoma, Crassatella, Phorus, Pectunculus* ...', 'Vol. 2. *Corbula, Arca, Triton* ...' all the way through to 'Vol. 20. *Solemya, Mya, Clausilia, Cylindrella* ...'. They aren't in alphabetical (or even taxonomic) order, but that would probably have been asking too much for a project that took more than 30 years to complete. I pull down volume three, '*Murex, Cyprae, Haliotis* ...' and carefully open the pages.

The beautiful colour illustrations are vivid and lifelike, almost as if the book were filled with real shells. The polished humps of cowries seem to perch on the page; an abalone bigger than my hand shimmers with many colours. This is the *Conchologia Iconica* by Lovell Augustus Reeve, one of Cuming's closest friends and associates. He began the book in 1843 and continued until his death in 1865, at which point George Brettingham Sowerby II took over, finishing the work in 1878. There are other splendid copies of the book in museums and libraries around the world and you can browse a digitised version online at the Biodiversity Heritage Library.

The *Iconica* series illustrates around 27,000 shells. Between the illustrations are written descriptions of each species, including many delightful common names. There's a

Greenish Cowrie and Yellowish Terebra, a Differently-bristled Bulimus and a Somewhat-distorted Triton; I spot a Grinning Cockle, an Ambiguous Murex, a Flea-bitten Cone, a Lovely Cone, a Melancholy Cone, and I can't help feeling a bit sorry for the Dismal Limpet.

The bulk of the shells that Reeve and Sowerby used for their drawings and descriptions came from Cuming's collection. The books include details about who found each particular shell and where. Repeated throughout, more often than any other name, is 'Mr Cuming'.

When Reeve started this great work, the Sowerbys had already begun to publish a five-volume shell guide, *Thesaurus Conchylorium*. I glance through a copy of this in the library and see why the *Iconica* was set apart as something quite different. The *Thesaurus* is illustrated with etchings, fine black lines painted in colour by hand. They are quite beautiful but give more of a stylised view of the shell rather than a realistic impression. Most of these drawings are also much smaller than the shells themselves. In contrast the *Iconica* illustrations are all life-size or bigger, which is one reason why it runs to 20 volumes (Reeve also wanted to include every known species at the time). And these are lithographs, a technique that allows for more subtle lines and shading. Until the invention of colour photography, this was probably the world's finest and most accurate book of shells.

The pictures in *Iconica* are so detailed and accurate it's possible to search through the museum's cabinets and find the actual, individual shell from the collection that Reeve or Sowerby drew. And I can't resist going back for one last look at Cuming's shells.

In a small side room off a long corridor, Jon directs me to the cabinets of cones and I open a few drawers before spotting what I'm after: *Conus gloriamaris*, the Glory of the Sea.

For a long time it was thought that the Glory of the Sea was extinct. For 60 years, from 1896 onwards, no specimens were found; it was obviously rare to begin with, and perhaps too many greedy collectors had exhausted the last wild stocks. Eventually, though, with the increase in bottom-dredging and the invention of scuba-diving, more specimens started showing up from deeper waters. Today, Glories of the Sea have become so common they aren't nearly as valuable as they once were. There are hundreds on offer at online auction sites, where you can snap one up for a fraction of their former price.

I pull open the drawer to reveal several specimens of *Conus gloriamaris*. It is the first time I've seen one myself and the first chance I've had to look up close at the intricate markings that Bard Ermentrout, George Oster and colleagues modelled on a computer. I browse through the shells until I find two that are rather smaller than the rest, only a few centimetres long. The handwritten note with them has the scrawled initials MC. As I turn them over in my hand I begin to understand how easy it is for objects to become highly revered – sacred even – because of the connections they can trace to a person, a place and a moment in time.

Even here at the Natural History Museum, an institution founded on science and reason, the curators know only too well there are certain objects that are quite simply special. Walk through the entrance into the main hall with its lofty ceilings and stained-glass windows and it has the same awe-inspiring feel as a great cathedral. At the top of the stairs, past the seated sculpture of Charles Darwin, is a small gallery. On display are 22 objects selected from the museum's 70 million specimens, all of them with wonderful stories to tell.

Among them are some of William Smith's ammonite fossils that helped him work out that rocks are layered through time; there is a nautilus shell intricately carved in Holland in the seventeenth century, which was one of 400,000 objects in Hans Sloane's collection that formed the basis of the British Museum; there is the skull of a lion that lived 700 years ago at the Tower of London as part of an exotic royal menagerie; and there are intricate glass models of phytoplankton and jellyfish made in the mid-nineteenth century by Leopold and Rudolf Blaschka, who pioneered a sculpting technique that has since been lost and forgotten. These aren't just any old fossils, shells, skulls and glass ornaments but things infused with history, human endeavour and ideas.

My own personal pick of treasures from the museum would include some of Hugh Cuming's shells (with a note pointing out just how many thousands more there are down in the basement). He may not have been a man who came up with great thoughts and theories of how the world works, but his limitless passion for one group of animals opened up an astonishing view of the natural world that no one had seen before.

The Glory of the Sea drawn by Reeve in volume one, plate six of the *Conchologica Iconica* was found by Cuming on 'Jacna Island of Bohol, Philippines (found on the reefs at low water)'. Reeve goes on to explain that he chose to illustrate a small specimen found by Cuming because of its especially rich markings. There was, Reeve wrote, another even smaller shell that Cuming collected that same day 'scarcely exceeding an inch and a half in length'. But he confessed the patterns on that one were so extremely fine they defied his drawing skills.

How strange it is to imagine that moment, 175 years ago, when Cuming stood on a beach in the tropical heat of the Philippines, lifted up a rock and for the first time saw the very shells that I'm holding now.

I wonder if finding them really did make him do a little dance.

CHAPTER NINE

Bright Ideas

For hundreds and thousands of years people have used molluscs and their shells as symbols of sex and death, as gems and ornaments and food, musical instruments and money, a source of golden fibres and things to simply gather together and look at. Now people are pondering molluscs and finding new and powerful things to use them for. And of all the mollusc species on the planet, it's the ones that do surprising things that are proving to be especially useful. There are molluscs that live in unlikely places, that are faster and stronger and more deadly than we might at first presume. Together they are giving rise to the next generation of mollusc-inspired ideas and discoveries. Some of the most ground-breaking innovations are coming from a group of slow, ponderous snails that hunt animals which should, by all rights, just swim off and leave them behind.

Unlocking the cone snails' secrets

During the daytime cone snails don't do much. They tuck themselves into crevices in coral reefs or bury themselves in sand, hiding away those intricate patterns drawn like memories across their shells. Only when dusk falls do these hunters emerge and begin their search for dinner.

There are approximately 700 species of cone snail, making *Conus* quite possibly the most diverse animal genus in the sea, and most of them have become specialist hunters that catch only one type of animal. Some hunt for worms, some for snails (including other cone snails) and some achieve the seemingly impossible: they eat fish.

After a cone snail has woken up, it will shuffle along, sweeping its proboscis through the water, probing for the scent of a sleeping fish. Silently, the predator picks up a trail and glides towards the oblivious target, then shoots out a hollow dart, loaded with venom. It impales the fish, paralysing it instantly. The snail then slowly draws in a thin cord attached to the dart and reels in its prey. Like a python swallowing an antelope, the snail distends its mouth to grotesque proportions, engulfing the fish and then settling down to digest its dinner. Several hours later the cone snail regurgitates a bundle of bones and scales.

Another tactic used by cone snails to catch fish involves doping them with puffs of sedatives that they release into the water. The snail will then unfurl its mouth into a huge net that draws in and smothers the snoozing prey, sometimes entire shoals of little fish at a time. Once they're bundled up inside, the snail shoots each one in turn with poison darts to make sure they don't try to escape.

The secret to the cone snails' expert hunting skills lies in their darts. These weapons are fashioned from individual, hollowed-out radula teeth with fearsome, backward-pointing barbs that get firmly stuck in the prey's skin. Each dart can be up to one centimetre long, and is only used once, like a disposable hypodermic needle. They are filled with venom and then stored, like a quiver of arrows, waiting to be deployed. When prey comes into range, a poison tooth is shot out at ballistic speed from the end of the proboscis by the squeeze of a sac called the venom bulb (the contracting muscle is assisted by the same enzyme that allows scallops and squid to achieve frantic bursts of speed).

As well as hunting small fish, cone snails can also kill people. They don't deliberately go after humans or consider us food; the deadly cones are simply defending themselves and will deploy their venomous darts if they feel threatened. Being picked up by an unwary fisherman or shell collector is enough to scare them, and because their bendy proboscis can reach around the entire body and shell there's no safe place to hold a cone snail.

When Dutch naturalist Georg Eberhard Rumphius was working for the Dutch East India Company in the early eighteenth century, he wrote about an Indonesian girl picking up a shell, feeling a 'tickling sensation' in her hand and then dropping dead on the spot. Since then there have been around 30 recorded cases of death by cone snail, mostly due to heart attacks and suffocation from the diaphragm being paralysed. The severity of a cone snail's sting depends on the species. Most won't actually kill you but their stings are unpleasant nonetheless, causing numbness and partial paralysis that can last for weeks. But whatever you do, don't mess with Geography Cone Snails: seven times out of ten, their sting is fatal to humans.

For such small animals to be loaded with enough venom to incapacitate a full-grown person is obviously quite over the top, and that is exactly why scientists became interested in them in the first place. For a long time, people have wanted to know why and how cone snails have become masters of chemistry and transformed themselves into such formidable killers.

The *why* part of that question is reasonably straightforward to answer. It all began with worm-eating cone snails, which were the first to evolve around 50 million years ago. It is thought these ancestral cone snails caught worms using relatively mild toxins, as many of the living worm-hunters do today. Before too long, though, they began to face competition from fish that sneaked up and tried to steal their dinner. The snails' response was to jab the intruders with painful stings to shoo them away. At first this was probably enough to deter the fish but it's possible they became more aggressive, and as a result the snails evolved more potent toxins. Eventually, their stings became so effective against intruding fish that it allowed the snails to switch to a whole new diet. In order to feed on fish, the snails needed toxins that instantly knocked them down; if the toxin takes even a few seconds to act, that would be enough time for the fish to swim off and collapse somewhere the snail may never find it. These powerful toxins are clearly a highly advantageous hunting tool; reconstructions of the cone snail family tree show that fish-hunting has evolved on at least three separate occasions. Time and again, as their prey got faster, the snails' venom became more toxic.

The bigger, more difficult question to answer about cone snails is precisely *how* their venom is so powerful. This is a conundrum that has kept researchers busy for decades. It was Alan Kohn at Yale University, back in 1956, who first saw cone snails hunting for fish and set out to understand how they detect their prey. He carefully put cones in aquarium tanks and watched as they buried themselves in the sand up to their eyes. Then he offered them various things to eat. A living fish dropped in the tank elicited an immediate hunting response; the cones would rise up out of the sand and start searching. In contrast, they totally ignored dead fish, but they were excited by a few drops of water from an aquarium in which living fish had been swimming; the snails set out hunting even though there was nothing for them to find. Through these experiments, Kohn had landed on the idea that cone snails sniff out prey. He went on to become a professor at the University of Washington and a world authority on cone snails, the animals that almost bear his name.

Investigations into the active ingredients of fish-hunting venoms began in the 1970s at the University of Queensland, where Bob Endean and co-workers had the Great Barrier Reef and a ready supply of cone snails on their doorstep. They were the first to figure out that the venoms are mixtures of several compounds. However, back then, no one yet suspected quite how elaborate the cone snail's toxins truly are.

The next major research efforts looking at cone snail venoms – which came to be known generally as conotoxins – took place in the 1980s in labs run by Baldomero Olivera. Originally from the Philippines, Olivera studied in the United States where he specialised in DNA and enzymes, and on returning to his home country found himself at a university with very little equipment to continue his molecular research. As a child he collected shells and knew very well the deadly reputation of cone snails, so he decided to test their venom on laboratory mice, a technique that didn't require much kit.

Olivera started with a simple experimental set-up that involved persuading mice to cling upside down to a horizontal wire screen, then injecting them with different extracts of cone snail venom (he split the venom into fractions of different-sized molecules). He then timed how long it took for each extract to take effect. When mice were paralysed, they would let go of the screen and fall off: this was the 'falling time'. Early studies like this showed that mice were paralysed by some venom extracts – but not all of them. Olivera's next goal was to see if the non-paralytic parts of the venom had any other effects.

A breakthrough took place back in the US, in Olivera's new lab at the University of Utah, where he enlisted the help of some smart undergraduates. One of them, Craig

Clark, came up with the idea of injecting venom extracts directly into the nervous systems of mice. Olivera admits that at the time he wasn't convinced this would work, but Clark carried on regardless. A succession of students perfected the technique and, before their eyes, mice started behaving very strangely. Depending on which venom extract was injected, the mice would tremble or scratch themselves uncontrollably; others would fall into a hypnotic trance for 24 hours then snap right out of it; and some would frantically run around their cages and climb up the walls.

It became clear that conotoxins affected different parts of the nervous system in very different ways. By the 1990s research groups around the world had caught on to the idea that cone snails and their venom held great potential for studying nerves and brains, and perhaps for developing new pharmaceuticals. Soon these snails became some of the most intimately studied animals from the oceans.

An enormous amount is now known about cone snails and their complex venoms. We know that conotoxins are composed of a mixture of peptides. Most are made of between 10 and 30 amino acids with lots of disulphide bonds that sculpt them into small, stiff shapes. We know that each cone snail species has its own signature mix of between 50 and 200 peptides, which they blend in their venom ducts. Along the length of the venom duct there are genes that are switched on and off, resulting in a tailored cocktail of peptides that trickles along the tube before being loaded into the hollow darts at the end. The cones can even adjust the recipe depending on whether they are hunting or stinging in defence.

It's not known exactly how many conotoxins there are. Each of the 700 known cone snail species produces its own unique blend of toxins, so it follows that there could easily be tens or even hundreds of thousands of conotoxins. No wonder there's no cone snail anti-venom, because it would need to counteract each individual peptide. Only a minute fraction of them have so far been identified and studied, but those that have are revealing the intricate molecular secrets of the cone snail's chemical armoury.

In general, conotoxins disrupt the passage of nerve impulses around the body by blocking or jamming neural signals. Normally, nerves fire when ions pass in and out of them to generate or dissipate electrical charges. The ions that carry these charges are sodium, calcium, chloride and potassium. Nerve membranes are densely dotted with channels made of protein that open and close to control the movement of each particular ion. These ion channels come in many varieties. The two most common types are those controlled by chemicals and those that respond to electrical charge itself, so as to either amplify or dampen ongoing activity. For a chemical to control an ion channel it must bind to a receptor on the channel and instruct it to open or close, much like a key in a lock. These chemicals include neurotransmitters that pass signals between nerve cells, allowing the brain to process information and communicate with the rest of the body. Together, all these ions, ion channels, receptors and signalling molecules govern the transmission of nerve impulses around the body, and many other complex cellular processes. When conotoxins come along they mess up this finely tuned symphony of ions by acting like signalling molecules, binding to ion channels and waywardly telling them to either open or close.

Based on their effects, different conotoxins have been sorted into groups that Baldomero Olivera and colleagues have nicknamed toxin cabals. Like secret societies plotting to overthrow a government, conotoxins gang up to overthrow the cone snail's prey. The toxin cabals were first uncovered in the Purple Cone Snail, a species which launches a two-pronged attack on fish. First they unleash the 'lightning strike cabal'. This makes nerves fire uncontrollably, essentially giving the victim a massive electric shock. It happens because of the combined effects of a conotoxin that jams open channels, causing an influx of sodium ions together with another that blocks potassium channels, preventing these ions from leaving. The upshot is a very still, very stiff fish.

This gives time for a second toxin cabal to kick into action. The 'motor cabal' blocks signals that pass between nerves and muscles; this takes slightly longer than the 'lightning strike cabal', because the conotoxins have to reach the ends of nerve fibres. Once the 'motor cabal' gets going it causes total and irreversible paralysis. Acting synergistically like this, the two conspiring cabals have everything covered, leaving little hope for a fish harpooned and reeled in by a Purple Cone.

Conotoxins are probably the most complex poisons on the planet. Other deadly creatures tend to rely on a single, albeit powerful toxin. In order to match the potency of cone snail poisons, and the intricate ways they affect living bodies, you would need to assemble a horde of other dangerous species. Not only would you need to lick the skin of a poison arrow frog (batrachotoxin) but also take a bite of liver from a pufferfish (tetrodotoxin), become infected by a colony of *Clostridium* bacteria (botulin) and to finish it off you would need to get bitten by a cobra (cobratoxin). Being deadly and biologically complex in their own various ways, many of these natural toxins have been used in biomedical research, but none have attracted quite as much attention as cone snail venom. The thing about cone snail venoms that gets neuroscientists really excited isn't so much the fact that they can kill fish and human beings, but more their exquisite specificity. Even though they are made from only a short string of amino acids, conotoxins are immensely picky about which ion channels they bind to.

There is a bewildering array of ion channels and receptors dispersed around an animal's nervous system, and each is a particular shape. A single conotoxin will only bind with one highly specific type of channel. They are very exact keys that only work in one particular lock. This makes conotoxins immensely powerful research tools. They allow neuroscientists to reach into a nervous system and choose precisely which components they want to switch on or off as they investigate the inner workings of nerves, brains and entire bodies.

In thousands of studies, conotoxins have aided researchers in understanding the fundamental processes of living things; they have advanced our understanding of how muscles contract, how blood pressure is regulated and how kidneys and retinas work. Conotoxins are showing just how many types of receptor there are and revealing the complexity of the human brain. Researchers are now figuring out the roles played by distinct receptors in neurological diseases like Parkinson's, Alzheimer's and alcoholism. And as well as helping to understand diseases, conotoxins are also helping to build a new arsenal of medicines to tackle them. For 50 million years cone snails have been evolving and perfecting the precision of their toxins; now biochemists are tapping into this immense repertoire and finding conotoxins that have specific therapeutic effects on the human nervous system. Dozens of conotoxin-inspired medicines are in development for treating an immense variety of disorders.

In the 1980s, Olivera's team found a conotoxin extract from the Geography Cone Snail that induced a sleep-like state in laboratory mice. This 'sleeper peptide' is one of the soporific drugs that net-hunting cone snails waft into the water to sedate their prey. The active ingredient was later identified as conantokin-G, a conotoxin in the 'nirvana cabal' that blocks a specific type of ion channel receptor for the neurotransmitter glutamate, called the NMDA receptor. Clinical trials are currently underway to see if this conotoxin could help calm the hyperactive nerves of people suffering from intractable epilepsy. It could also stop the breakdown of nerves in patients with Alzheimer's and Parkinson's. Other conotoxins are being investigated as treatments for heart attacks, multiple sclerosis and ADHD. And for more than a decade now, people suffering from chronic pain have been deliberately injected with cone snail stings.

Ziconotide, marketed as Prialt (the 'primary alternative' to morphine), is an artificial version of a conotoxin originally found in the Magician's Cone Snail. Prialt blocks calcium channels that transmit pain signals from nerves to the spinal cord and on up to the brain. It is a thousand times more potent than morphine, with much lower risks of addiction. The main drawback, though, is that it must be injected directly into the spinal fluid using a small pump inserted under the skin, an obviously intrusive procedure. A research team at the University of Queensland, where Bob Endean's early studies of cone snails were carried out, are now working on conotoxin pills. To do so, David Craik and his colleagues are making and testing synthetic conotoxins that are looped around into rings, making them more stable and likely to survive passage through the human digestive tract.

And it's not just conotoxins that are inspiring new drugs from cone snails. It turns out their weaponry is even more complex than previously thought. In 2015, a startling new finding emerged from a team at the University of Utah that included Baldomero Olivera. The study, led by Helena Safavi-Hemami, revealed that some cone snails send fish to sleep using sedatives laced with insulin. The peptide hormone elicits hypoglycaemic shock – a dangerous drop in blood sugar levels – making the fish pass out. Various forms of insulin are components of the 'nirvana cabal' of the Geography Cone and Tulip Cone, and they have evolved to be structurally more akin to fish hormones than molluscan varieties. These are the first known cases of weaponised insulin, and they open new avenues of research into how insulin works and, while it's still a way off, there's the potential for developing new drugs to treat diabetes.

Researchers have come a long way from watching sea snails master the unlikely skill of fish hunting. With so much knowledge generated and so many new ideas for medicines, the most astonishing thing of all is to contemplate what we still don't know. There is so much left to discover inside cone snails and their tiny harpoons. So far, only around a hundred conotoxins have been studied in detail from six species, which surely means these masters of chemistry still have a thing or two left to teach us.

Sticking, gluing and digging

Down at the coast there are seashells that sit tight in some most improbable places. Mussels glue themselves to wet, slippery rocks despite the relentless crashing and sucking of waves around them, and for decades scientists have watched in envy, desperate to find out how they do it. Underwater superglue is another of the latest inspirations from molluscs.

Herbert Waite was one of the first people to begin unlocking the mussel's sticky secret. As a graduate student at Harvard in the 1970s, he began gathering mussels from the northern shores of Long Island Sound in Connecticut, on the US Atlantic coast. Back in the lab, he scrutinised the byssus fibres that mussels use to anchor themselves, and Sardinian weavers use to spin golden threads. He broke the proteins down into their separate components, and among them he discovered a rare amino acid called L-dopa.

L-dopa had been known of for a while. It is present in various plants and animals and the human body makes it as a precursor to the neurotransmitter dopamine. For this reason it was developed as a treatment for Parkinson's disease and other conditions. The 1990 movie *Awakenings* tells the true story of how neurologist Oliver Sacks used L-dopa to rouse patients from decades of catatonia.

Waite was the first to pinpoint the role of L-dopa in mussel glue. He worked out that this amino acid is key in allowing the liquid protein, secreted by the mussel's byssal gland, to set hard in saltwater and stick the mussel in place. Since his discovery, many different glue proteins – referred to as Mussel

Adhesive Proteins or MAPs – have been identified, and they all contain this molecule.

Currently researchers, including Herbert Waite and his lab now at University of California, Santa Barbara, are studying how MAPs work. The full picture has yet to be revealed but an important factor has been established. L-dopa molecules contain side-chains called catechols that interact directly with surfaces, be it a rock, a boat hull or whatever a mussel is trying to stick to, forming bonds that fix the mussels in place.

Synthetic glues laced with L-dopa or other catechol-containing compounds are being developed, with lots of potential applications. Most immediately, mussel-inspired glues are likely to be used inside the human body. A glue that works in blood vessels, with blood coursing through them, will be incredibly useful to surgeons. In particular, foetal membranes are very difficult to repair and bio-glues are being tested as a suture-free way of operating on unborn babies.

In addition, patients suffering from atherosclerosis and the build-up of plaque inside their blood vessels could be treated with a squirt of glue in their arteries, to help prevent heart attacks and strokes. Currently, stents or balloon angioplasties inserted into blood vessels to widen them are smeared in anti-inflammatory drugs, but around 95 per cent of the drug gets washed away in the blood flow. Bio-glue could see an end to this wastage.

Diabetes could also one day be treated with a dab of mussel-inspired glue. Instead of having to inject insulin, an alternative is for diabetics to have pancreatic cells from donors transplanted inside their bodies to produce insulin on their behalf. Currently, it's possible to insert these cells inside the liver, but they only work for a few years. Using bio-glue, it might be possible to find somewhere else to stick these cells, on the outside of the liver, perhaps, without triggering inflammation and giving them a much longer lifespan.

One of the latest ideas to emerge from Herbert Waite's lab is a synthetic polymer, covered in catechol-rich proteins, that can heal itself. Potential applications for this new material include the manufacture of hip and knee replacements that would require little or no surgery to maintain. It could also be used to reinforce hairline cracks in brittle bones. The mussel-inspired polymers could even one day be used to make self-repairing surfboards.

Beyond the human body, and wave riders, there is another, rather unexpected use of mollusc glues. MAPs could be used to *stop* molluscs from sticking. Fouling organisms are the weeds of the sea, growing in places they aren't wanted. When molluscs and barnacles clamp on to boat hulls they increase drag in the water, pushing up fuel bills. And while many boats these days are metal or fibreglass, shipworms and their boring habits are still a threat to wooden jetties and pontoons.

Various treatments have been developed over the years to try to stop these nuisance creatures from getting a grip, but one in particular turned out to be grimly toxic. Tributyl tin, or TBT, was banned worldwide in 2008 after it was found to cause all sorts of ecological problems when used as anti-fouling paint (TBT compounds deter marine larvae from settling on treated surfaces). Alarms were raised when marine biologist Stephen Blaber found that female dog whelks around the British coasts were sprouting male genitalia. Far from being a minor inconvenience, a female whelk exposed to trace amounts of TBT will sprout a penis so large it blocks her oviducts, preventing eggs from being released and rendering her infertile (ecologists now routinely measure the length of wild female whelk penises as a gauge of environmental pollution). Maritime industries are still hunting for replacements for TBTs; one possibility is to use mussel glues to stick other, less harmful anti-fouling agents firmly to boats to keep their bottoms clean.

While mussels use chemistry to spend their lives stuck implausibly to rocks, another group of bivalves have become masters of physics and move in a way that at first seems impossible. Razor clams are long, narrow bivalves that spend much of their lives buried in sandy, muddy shores. They dig using a two-anchor system, opening their twinned shell slightly to hold it fast while pushing their muscly foot into the sediment. The clam then pumps blood into its foot, making it swell up and act as a second anchor while it pulls the shell downwards.

Based on the shape, size and strength of razor clams, calculations indicate that they should only be able to dig a short way before getting stuck by the pressure of mud and sand crushing down on them. Researchers have tested this by shoving model razor clams into a sandy beach. Test shells only penetrated a couple of centimetres (about an inch) beneath the surface. By contrast, the Atlantic Jackknife Clam, which measures around 20 centimetres (eight inches), can dig far deeper than its muscles and shell alone should allow. Razor clams have evolved a way of being so energy efficient they could burrow half a kilometre (a third of a mile) using just the power in a household AA battery. It turns out that the key to their digging skills lies in a little puddle of quicksand. Repeatedly opening and shutting their shells causes hard sediment around a razor clam to collapse, and water seeps inwards, creating a pocket of liquidised sand or mud. This reduces drag and cuts down the energy needed to burrow by around 10 times.

A robotic version of a razor clam is helping Amos Winter to explore this idea of quicksand digging. Along with his colleagues at the Global Engineering and Research Lab at MIT, Winter has spent a lot of time wading through mudflats in Gloucester, Massachusetts with RoboClam in tow.

Winter trained RoboClam to be as good at digging as possible by using an algorithm inspired by natural selection. This approach mimics biological evolution by generating hundreds of random tweaks in the design of little piston-powered clams, and testing which works best. The result is an imitation clam that can dig as effectively as a real one. Winter envisages that RoboClam will one day lead to small, low-power digging systems that could drastically cut the costs of anchoring boats. Even more excitingly, engineers have their eyes on international internet traffic, now that almost all this information passes over the seabed along submerged communication highways (a far cheaper method than using satellites). RoboClam's descendants might one day be used to pin down the fibre-optic cables that reach between continents and wire up the Earth.

Their ability to stir up pools of quicksand is not the only thing that makes razor clams such expert diggers. They also depend on their shells not snapping in two. And like all molluscs, their shells are surprisingly strong.

Shiny on the inside

Being essentially made of chalk, seashells should be easily shattered. If you've ever snapped a stick of chalk in two you'll know what I mean. And yet you can squeeze them, drop them, hit them with hammers, do what you like (within reason), and many mollusc shells will stay in one piece. This poses yet another conundrum that endlessly teases scientists and engineers. Why don't seashells break all the time? Why don't they snap as easily as a stick of chalk?

Hunting for answers to these questions has led researchers into the inner structure of seashells and the nano-world of the incredibly small. At this scale molluscs have evolved elegant solutions to the problems of attack, and found ways of not getting smashed. And it's here we find inspiration for constructing protective shells of our own.

In the last few years there's been a surge of interest among scientists and engineers in mother-of-pearl. This is the shiny layer on the inside of shells, also known as nacre, and the stuff that pearls are made of. It's well known that pearls are the result of parasites or bits of grit that irritate a bivalve's soft innards; the mollusc envelops these foreign bodies in layers of smooth nacre to protect itself (a fact that gem merchants rarely admit). Material scientists are busy learning lessons from nacre; not, though, on how to make things iridescent and pretty, but on how to make things incredibly strong.

Adult mollusc shells are commonly made of aragonite, a tough form of calcium carbonate that offers good protection against crab claws, fish jaws and even attacks from its own kind as they try to drill and stab their way in. However, the layer of calcium carbonate is prone to fractures. When cracks appear on the outside of a shell they spread, but they stop as soon as they reach the shiny inner layer of nacre. This happens because of nacre's unique architecture.

Seen through an electron microscope, nacre is composed of diamond-shaped crystals, stacked like bricks in layers on top of each other, made of 95 per cent aragonite. At around 300 to 500 nanometres thick, these layers are just the right size to make waves of visible light bounce around between them, creating the structural colours that give mother-of-pearl its gleam. These nacreous bricks are mortared together with much thinner layers of chitin, the protein component of insect and crustacean exoskeletons. When cracks spread through nacre they are sent on a tortuous path between those microscopic bricks, which saps their energy and stops them in their tracks. The bricks slide over each other and the chitin protein stretches, helping to dampen the growing fracture.

Layers of nacre are also somewhat kinked, which seems to make them even more crack resistant. Mohammad Mirkhalaf and François Barthelat from McGill University demonstrated the importance of this waviness in nacre by mimicking its nano-scale structure. They used a laser to etch wavy lines into small glass plates and showed, counter-intuitively, that the glass became tougher by scratching it.

Over recent years several research teams have used a range of ingredients to make synthetic nacre. A team from Manchester and Leeds Universities used calcium carbonate mixed with polystyrene to make crack-resistant ceramics that could one day be used in building materials and bone replacements. Lorenz Bonderer and a team at ETH Zurich mimicked nacre using thin plates of aluminium oxide coated in chitosan (a material derived from shrimp and crab shells with added sodium hydroxide) to make a composite material that could be used to make aeroplanes and spacecraft. And in 2012, a group from Cambridge University were the first to make artificial nacre the way nature intended. Alex Finnemore and Ulli Steiner headed up a team that imitated the steps molluscs themselves take to produce nacre. They laid down layers of calcium carbonate and then sandwiched them between protein sheets pitted with pores. This artificial nacre looks, feels and behaves just like the real thing.

The next, eagerly awaited step is to start using mollusc-inspired materials for applications in the human world. In the pipeline are visors for motorcycle and space helmets, based on a mollusc named, appropriately enough, the Windowpane Oyster. These oysters, also known as capiz shells, were traditionally used in Asia to make windows, and more recently have been made into all sorts of decorations; a roaring export trade in lampshades, candle holders and Christmas lanterns called paról has wiped out stocks of these molluscs in the Philippines. The shells are not just decorative and see-through – they also happen to be incredibly strong.

Ling Li and Christine Ortiz from MIT worked out that Windowpane Oysters, which are 99 per cent calcite, have a similar nanostructure to nacre, with layers of elongated hexagonal crystals. In the lab, Li and Ortiz whacked chunks of these shells with a diamond-tipped hammer, then inspected the damage under an electron microscope. They saw the crystals behaving in various complicated ways including so-called nanocracking, visco-plastic stretching and nanograin formation; suffice to say that at a nano scale, the crystals put a halt to the spread of damage and set up a no-go buffer zone that cracks don't cross. It means that unlike artificial ceramics, damaged shells stay largely intact and crystal clear. So, if their oyster-inspired helmets get cracked, astronauts will still be able to see out.

Composite ceramics, based on ideas borrowed from deep-sea snails, could one day show up in military body armour and vehicles. The Scaly-foot Snail was discovered on the Kairei hydrothermal vent field in the middle of the Indian Ocean, more than 2,000 metres (1.2 miles) beneath the waves. The scaly-foot gets its name from its strange covering that looks rather like the sclerites that covered the Cambrian creature *Wiwaxia*. That would seem odd enough for a modern mollusc, but even more bizarrely their scales are made of *iron*, and their shells are too. No other organisms are known to make skeletal structures clad in iron.

The Scaly-foot Snails make tri-layered shells that trump those of all the other molluscs. Sandwiched between hard layers of iron sulphide on the outside and calcium carbonate on the inside lies a spongy organic sheet. Back in Christine Ortiz's lab at MIT, her research team tested the protective capabilities of these peculiar iron shells, again by bashing them with sharp probes, then bathing them in hot acid and programming computer simulations of predator attack. They worked out that each layer in the shell has its own distinctive role in protecting the squashable snail inside.

The inner calcium carbonate layer, as in all molluscs, provides an unbending scaffold that is strong but prone to fractures. The organic mid-layer is padding that dulls the blow from attacks; in the wild these snails are hunted by crabs that grab hold of them and can keep on squeezing for days. In addition, the organic layer protects the inner shell from overheating and corroding in the scorching, acidic waters that gush up through the hydrothermal vents. The outer covering of iron sulphide (in fact a form of the compound called greigite) has a nano-scale structure that, similar to nacre, stops cracks from spreading through the shell; it probably also blunts the claws of crabs that try to smash their way in.

The iron-rich scales that give Scaly-foot Snails their name help them to survive attacks from another mollusc species that inhabits the same deep sea vents. Turrid snails hunt in a similar way to their close relatives, the cone snails, firing out venomous darts. The Scaly-foot Snails protect themselves from the rain of arrows by cladding their feet in chain-mail armour. Compared to the cones, very little is known about the venom of turrid snails, and within these minute molluscs even greater pharmaceutical treasures may await discovery.

In 2008, Baldomera Olivera went searching for microsnails in the Philippines. He worked with a big team, including Romell Seronay, who tested a highly effective but simple collecting tool: two armfuls of knotted, broken fishing nets. These were tied to a weight and lowered 40 metres (130 feet) down into the clear waters off Balicasag Island in the central Philippines and left there for six months.

Known as *lumun-lumun*, this fishing technique was developed in the Philippines to meet a highly unusual demand. There are shell collectors, mainly in Japan, who devote their spare time to gazing at teeny tiny shells down a microscope. Fishermen worked out that placing their old nets in certain areas of the sea was an ideal way of gathering up these diminutive molluscs. The fine netting acts as temporary habitat for drifting mollusc larvae, which settle down and start growing. Other small but mature molluscs will creep in and seek refuge in the tangled mesh.

After waiting patiently for months, Seronay and the team hauled in and shook their net bundles, and got quite a surprise. Out dropped more than 200 mollusc morpho-species – that is unidentified, probable species. The haul included five new cone snail species and 30 turrids, all of them smaller than half a centimetre long.

The team dissected out the venom ducts (a fiddly job) of the most common turrid in their catch, a tiny thing called

Clathurella cincta. Sequencing the DNA from *Clathurella's* venom duct, they found genes for two novel peptides similar to conotoxins and presumably with some form of neurotoxic effect. This small project was proof of the concept that *lumun-lumun* fishing could open up a whole new window onto the pharmacological treasures of the deep.

Cone and turrid snails, super-strong nacre, iron-clad deep sea snails and sticky mussel glue together make a compelling case for protecting marine life. Even if it's for no reason other than self-interest we should care about keeping ocean ecosystems as healthy and intact as possible, just in case there are more things out there that will one day be useful in solving human problems.

There is, however, a potential paradox in this argument. What if too many people want to get their hands on these useful species? For cone snails in particular, there is widespread concern that they are being taken from the wild in vast numbers to feed a growing demand from research labs around the world.

In the past, the only way to get hold of conotoxins for research was to grab a living cone snail (very carefully) and chop out its venom duct. Fishermen in the tropics came to specialise in catching cone snails for this very purpose. The exact volume of the trade is unknown, but a US laboratory reported buying consignments of venom ducts a kilogram at a time. Each kilogram would have contained the ducts from around 10,000 snails.

Since then, techniques to keep cone snails alive in captivity and milk their venom have been developed, but this is not for the faint-hearted. One of Olivera's students was the first person to rub an inflated condom on a goldfish, then offer it to a cone snail. The snail dutifully obliged, launching an attack, and seconds later the condom was bobbing at the surface with a poison dart lodged in it and the snail dangling down. More recently, advances in sequencing technologies, the ability to amplify DNA from tiny samples and to make peptides in the lab should see an end to the great piles of dismembered cone snail ducts. Still, though, cone snails face many other threats.

In 2013, a global assessment of 632 cone snail species revealed some key facts about their status in the wild. On the one hand, around three-quarters of all cone snail species seem to be doing reasonably well; they are widespread and abundant enough that they aren't at risk of going extinct anytime soon. A question mark hovers over 87 species that haven't been assessed due to lack of data. The remaining 67 cones – around one in ten known species – are considered to be at risk of extinction or likely to head that way in the near future. If we are to maintain the option of studying and using those cone snails and their complex conotoxins, all these species need protecting.

One reason for their threatened status is that many cone snail species have highly restricted ranges. There are species that are found only in the waters around one island or even in just a single bay. As was suggested for the extinct ammonites, the species with smaller ranges are often more likely to go extinct, especially when their habitat is at risk. The stories are sadly familiar. Two species of cone snails found only in Florida are losing their habitat to condominiums and tourist resorts; several Caribbean islands, including the Bahamas, Martinique and Aruba, have their own unique cone snail species, and these are at risk from collectors taking too many. The majority of the world's endangered cone snails live in the eastern Atlantic, in the Cape Verde archipelago and on the coast of Senegal around the capital city, Dakar. These snails are at great risk from sprawling coastal development and encroaching urban pollution.

With so many endemic cone snails living in small areas of habitat, the spotlight falls on local conservation efforts; the future of each of these species will depend on what happens at a local or a national scale. Meanwhile, another threat is looming on the horizon for all the cone snails, and the rest of the marine molluscs across the world, one that will need a global solution. The shifting composition of the atmosphere due to humanity's carbon emissions means that some seashells could soon begin to simply melt away.

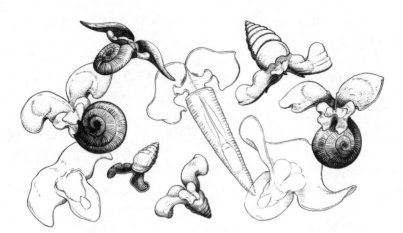

CHAPTER TEN

The Sea Butterfly Effect

A sea butterfly flutters past. Its spiralling shell is translucent and colourless as though it were sculpted from glass. Inside, I see a cluster of cells twitching and contracting as its heart beats. Little wings stick out from the shell's flared opening and flicker in energetic bursts, propelling it through the water in circles. It stops now and then as if to catch its breath, and I hold my own as I quietly watch, partly so as not to disturb it but also because this is the first sea butterfly I've seen and I can't quite believe my eyes.

It quivers one more time and flits out of sight. I sit up and look at the shallow petri dish on the laboratory bench in front of me. I can just make out a tiny, whirling dot and suddenly feel like I've been Alice in Wonderland, peering through a tiny door into another world.

Earlier that morning, the sea butterfly had been swimming through clear, deep waters that surround the island of Gran Canaria. This parched volcanic outcrop lies 100 kilometres (60 miles) west of mainland Africa, at the same latitude as the desert border between Morocco and Western Sahara. I had come to meet Silke Lischka, a sea butterfly expert who had kindly agreed to help me find one of these beautiful, peculiar molluscs that could easily have sprung from the imagination of a storyteller. I desperately wanted to see one for myself, to check that they are real. And I wanted to see them now because their time might be running out. These fragile animals could one day soon begin to vanish from the seas, the early victims of climate change and a silent warning of troubles to come.

We had motored offshore on a black, inflatable research boat across the sea, flat like a swimming pool and only ruffled here and there by a gentle breeze. We found a good spot, stopped the engine, and Silke then lowered a plankton sampler into the blue water. Peeping over the side, I watched the rope paying out 15 metres (50 feet) or more, visible all the way as it dragged the white net down like a slender, upside-down parachute. On its return journey back to the surface the net sifted seawater, trapping anything bigger than a fine sand grain (70 microns, or 0.07 millimetres). Hauling the net back on board, Silke carefully unclipped the canister that had caught the siftings and tipped the contents, about half a litre of water, into a small screw-top barrel. I looked in and saw a blizzard of swirling particles, and immediately started imagining what we might have caught.

Six or seven times, Silke plunged the net down then dragged it back up, bringing in more minuscule treasures until she decided that we had enough to be getting on with. We kicked the engine into life and returned to land, passing flying fish that skittered through dry air on their improbable wings before plopping back down to where they usually belong.

Back in the laboratory at PLOCAN, the Plataforma Oceánica de Canarias, we sat diligently working through the plankton samples, pouring out small pools of seawater and examining their contents through microscopes with up to 40 times magnification. We had captured a fidgeting, living galaxy. There were masses of minute crustaceans called copepods, with bodies shaped like tear drops and some with a single, red cyclopsian eye; they paddled through the water on pairs of long whiskery appendages and turned endless pirouettes, chasing their tails round and round. Fuzzy tufts of cyanobacteria, or blue-green algae, drifted past like tumbleweed. I spied some *Noctiluca scintillans*. Under the microscope these dinoflagellates (a type of green algae) look like transparent peaches. At night, in their millions, they transform the seas into a glittering light show of biolumi-nescence. There was a tunicate larva with a small head and wriggling tail; how strange to think that, in time, it would settle onto the seabed, absorb its brain and become a plant-like sea squirt. I saw radiolarians like exquisite, many-pointed stars, pulsing cuboid jellyfish larvae, and foraminifera with coiled, chambered bodies that could be mistaken for miniature ammonites. But most splendid of all, I was quite convinced, were the gastropods with tiny wings.

For a while we saw no sea butterflies and I began to worry that I'd missed my chance, that it was too late in the season and the Atlantic had already become too cold and empty of food for them to still be hanging around. But we carried on, in hushed concentration, working our way through the barrel of seawater, until eventually Silke let out a little giggle and told me to come and take a look. She had found a small specimen of *Limacina inflata* (sea butterflies tend not to go by common names, only their scientific labels). Her find seemed to break the spell of the hiding sea butterflies, and suddenly plenty more showed themselves. Silke spotted a different species, not with a spiralling shell but with a delicate, conical tube instead. I began to get my eye in and found a sea butterfly for myself and it felt all the more special. I was the first person ever to lay eyes on that particular tiny creature.

'They look like little snitches,' said Silke, chuckling. And they do. When J. K. Rowling created the game of quidditch, played on broomsticks by the pupils at Hogwarts School of Witchcraft and Wizardry, and the small golden ball with wings (which Harry Potter caught many times and swallowed at least once), I'd like to think she was inspired by sea butterflies. I watched them, transfixed, as they spun around, busily inspecting their shrunken sea as if they had somewhere important to get to. Soon, I became convinced that I was a natural-born sea butterfly-spotter. I spied sea butterfly larvae, which are so much smaller than the adults. Side by side they were pea and grapefruit. The young ones haven't yet grown wings but have two lobes that are covered in tiny wriggling hairs and whir in circles, like an industrial floor-polisher. The movements of these energetic adolescents made the water around them glimmer and dance in a certain way that I learned to recognise and zero in on. And I found another minute mollusc with a spiralling shell that looked similar to the rest but with one important difference. I showed it to Silke and she raised her eyebrows at me, smiling; I knew I had earned brownie points. It was a heteropod, a distant relative of sea butterflies from another deep division of the gastropods. Unlike the sinistral sea butterflies, this one had a shell that twirled to the right.

Sea butterflies are also known as pteropods, the 'wing feet' creatures (just as pterosaurs were 'winged lizards'). These most unlikely gastropods have wings instead of feet, which they use to swim through open seas worldwide, occupying the biggest living space on the planet. They are perhaps the most abundant animals that almost nobody has heard of.

Other pteropods, known as sea angels, also fly about underwater, but these have lost their shells. Instead, to protect themselves, their bodies are loaded with noxious chemicals that attackers soon learn to avoid. Their chemical defence is so effective that small crustaceans called amphipods have learned to kidnap sea angels and carry them around, keeping them alive, like personal bodyguards. However, don't be fooled by the angelic appearance of the sea butterflies' shell-less relatives. Sea angels are compulsive predators that hunt exclusively for sea butterflies. They have keen eyesight to spot their prey, fast wings to pursue them, and suckered tentacles to grab them and wrench them out of their shells in a violent battle of angels and butterflies.

Sea butterflies themselves get their food in an altogether gentler fashion. They cast webs made of sticky mucus and – just like spiders – they trap their food. Among the things that often wind up in their nets are crustacean and gastropod larvae (including of their own kind), phytoplankton, and obscure, vase-shaped animals called tintinnids. When it's ready, the sea butterfly hauls in the whole lot, eating its dinner, web and all.

Their gossamer webs are difficult to see but in the 1970s and '80s, two dedicated sea butterfly researchers found a way. Ronald Gilmer and Richard Harbison from Woods Hole Oceanographic Institution in Massachusetts spent a lot of time scuba-diving all over the world, tracking down these minute creatures and observing what they get up to in their natural habitat (many sea butterfly species grow large enough as adults to be seen with the naked eye). They would take a bottle of crimson dye with them and squirt drops into the water near sea butterflies to illuminate their webs. The animals would cast their nets and then hang motionless in the water – neither rising nor falling – giving Gilmer and Harbison the idea that sea butterflies might use their feeding apparatus to help them stay afloat, rather like the way that female argonauts use their shells.

Sneaking up and gently nudging them, the divers witnessed the sea butterflies' escape response: they quickly jettison their web, then either flit angrily away or pull in their wings and drop into the depths. Sea butterflies are good swimmers but they use up a lot of energy in the process. Many are negatively buoyant, and have to keep swimming or they sink. There are clearly benefits to be had from floaty nets, like tiny parachutes, that give them a break from all the incessant flitting.

There's a lot we still don't know about how sea butterflies move around their open ocean world. Silke shows me a video she shot of an Arctic species drifting through a large glass jar. She gently stirs the water and the sea butterfly stops beating its wings, holds them stiffly above its head and seems to ride the currents like a hawk on a thermal.

Sea butterfly procreation is especially curious. In some species there are separate males and females that will pair up, grab hold of each other's shells, and swim together in spirals through the water for a minute or two while the male transfers sperm to the female. She will then lay strings of fertilised eggs, which she may carry around with her, stuck to her shell, before the young hatch and swim off. Meanwhile, some species are sequential hermaphrodites; they all start life as males then later switch sexes, becoming females. Early in the spawning season, when there are only male sea butterflies, they will mate with each other. Males undertake a mutual sperm exchange, then hold on to their partner's donation until they turn into females. Then, all the new female need do is to fertilise her eggs using the donated sperm she's saved up from her earlier, male-only encounter. It might initially seem like an odd way of doing things, but it makes sense in the big, wide open ocean where finding a partner of the right species *and* the opposite sex can be difficult: by swapping genders and having sex in this unusual way, the sea butterflies increase their odds of finding a suitable mate.

Pteropods are not the only gastropods that have abandoned the sea floor. *Janthina* is a genus of snails with vivid purple, spiralling shells that float on the sea surface, buoyed up by a raft of frothy bubbles. *Glaucus atlanticus*, known as the sea swallow, is a shell-less gastropod that also occupies this two-dimensional world; it hangs upside down from the surface rather like a water boatman in a pond, with long fingerlike projections that store stinging cells scavenged from its favourite food, the Portuguese Man-of-war. Spanish Dancers, another no-shell gastropod, can usually be spotted crawling across coral reefs but occasionally they fling themselves into the water and swim along with flamboyant ripples of their mantle, like a flamenco dancer's twirling skirts.

All of these gastropods are drifting and swimming through seas that are silently changing and many of them – especially the sea butterflies with their tiny, fragile shells – could soon find their world turning sour.

Silke Lischka had come to Gran Canaria not to show me sea butterflies but to take part in a major, two-month research expedition, designed to help us understand more about what the future holds for these delicate molluscs and other minute sea creatures. Despite the gruelling work schedule, Silke had devoted her well-earned day off to helping me in my search, but she had to get back to studying what happens to sea butterflies when their watery world is threatened.

The problem of pH

For a little over two centuries people have been digging out and pumping up ancient black stuff from deep underground and using it to produce heat and light and food and to propel themselves about the place at ever-faster speeds. Burning all this coal and oil sends carbon dioxide in colossal quantities into the air where, together with other pollutants, it insulates the planet like a blanket, trapping the sun's radiation and leading to the various complex effects of anthropogenic climate change. But not all the so-called greenhouse gases released from burning fossil fuels stay in the atmosphere. Around a third of all the carbon dioxide ever made by human activities has been absorbed into the oceans.

If it wasn't for the saltwater that covers seven-tenths of the planet, the problems caused by climate change would already be unspeakably worse than they are today. Every hour, the oceans absorb a million tonnes of carbon. In less than four hours they absorb the equivalent of the annual carbon emissions from a coal-burning power station. We all have a lot to thank the oceans for.

The problem is that carbon dioxide doesn't just sit unnoticed in the oceans, but it has its own particular effect. When it reacts with seawater, carbon dioxide lowers the pH, making the oceans more acidic. Measurements show that since the dawn of the industrial revolution, ocean pH has fallen by 30 per cent. If we carry on with business as usual and do nothing to cut carbon emissions, experts confidently predict that by the end of the century ocean pH will have dropped by 150 per cent. There's no question about it: this is purely a case of indisputable chemistry.

The term 'ocean acidification' first became popular in 2003, when Ken Caldeira and Michael Wickett published a paper in the journal *Nature*. They calculated that if we go ahead and burn all the remaining fossil fuels, the oceans will become more acidic than they've ever been in the past 300 million years. Whether things will ever get that bad we'll see, but the point is that the chemistry of the seas is already changing.

Ocean acidification gets far less attention in the public eye compared to other threats linked to climate change. All we tend to hear about are rising temperatures and rising sea levels. Nevertheless, away from the media spotlight, researchers are beginning to untangle an important, difficult question: how will marine life react to acidifying oceans?

The fact is that the seas aren't exactly transforming into a caustic acid bath that would strip your skin off when you jump in. Surface waters of the ocean are still mildly alkaline, with an average pH of 8.1, compared to pH 8.2 200 years ago (the pH scale is logarithmic, which is why a drop from pH 8.2 to 8.1 equates to a 30 per cent change). Pure water has a pH of around 7; acids are below that, with milk at pH 6.5, lemon juice at pH 2 and stomach acid at pH 1. At the other end of the scale are strong alkalis, like household bleach with a pH over 12.

By 2100, average ocean pH could be down to 7.8, which is not exactly stomach acid, but in fact around the same pH as human blood. However, many marine organisms are adapted to living in water that has a fairly constant pH. Even a minor tweak to seawater pH could be enough to throw all sorts of things out of whack.

Laboratory studies on the effects of falling pH on marine life are producing plenty of findings, some of them rather unexpected. In water with carbon dioxide bubbled through it, young clown fish lose their sense of smell and become deaf, making it more likely that in the real world they would blunder into a predator or have trouble sniffing and hearing their way home to a coral reef. If the movie *Finding Nemo* had been set in the future, the little clown fish would probably have stayed permanently lost (or been eaten). Other fish species could become more anxious as the seas' pH drops. Californian rockfish kept in tanks of more acidic water became uncharacteristically shy, spending much of their time lurking in darkened areas and staying away from the light.

Ocean acidification could also make the seas more toxic in other ways. Lugworms live burrowed into sandy and muddy shores across northern Europe, where they are important food for wading birds and fish. You won't often see them, but they leave distinctive worm casts across beaches at low tide, like squeezes of sandy toothpaste. Recent studies show that copper, a common contaminant of coastal waters, is much more toxic to lugworms in acidified seawater. When pH drops, copper kills lugworm larvae and damages the DNA in lugworm sperm, making them swim more slowly and reducing their chances of reaching a fertile egg and forming an embryo.

These various subtle effects on behaviour and toxicity are difficult to predict, and researchers have to work backwards, unpicking the story and figuring out why these changes take place. However, for one particular group of marine species the effects of ocean acidification are much more foreseeable.

Calcifiers are a mixed gathering of marine organisms that all produce calcium carbonate in some form, as exoskeletons or shells. There are calcifiers stationed all the way through marine food webs, from microscopic, sun-fixing plankton, to sea urchins, starfish and corals, crustaceans and worms, and of course all those molluscs with shells. And these carbonate-makers are all in the firing line of ocean acidification.

The calcifiers' problems begin with the fact that calcium carbonate dissolves in acid. If you place a chicken's egg (also made of calcium carbonate) in a glass of vinegar you'll see this happening for yourself, albeit to an extreme degree: the shell dissolves leaving a naked egg, held together by a thin membrane. This sort of approach is the only way, so far, that anyone has investigated how acidifying oceans might affect argonauts. When Jeanne Power studied argonauts in Sicily in the early years of the industrial revolution, she had no reason to think of testing the effect of pH on pieces of their shells. When Kennedy Wolfe at the University of Sydney, Australia tried it in 2013, he found that at pH 7.8, argonaut shell begins to dissolve. This arises from the fact that female argonauts make their shells from an especially fragile form of carbonate, called high-magnesium calcite, which readily dissolves at lower pH. Argonaut shells also lack an outer, organic layer that could help protect other molluscs from acid attack (a spongy, thick protein layer is one reason molluscs can survive the corrosive conditions at hydrothermal vents).

What we don't know is how argonauts might react to falling pH while they are still alive (the animals are too rare and difficult to keep in captivity, so no one has tried this). Jeanne watched her animals use web-like membranes to fix damaged shells. Would argonauts do the same thing if their shells began thinning and dissolving in acidifying waters? Perhaps, but there is an added problem. As well as making their shells more likely to dissolve, ocean acidification also makes it harder for molluscs to make and mend their shells.

When carbon dioxide reacts with water it not only releases hydrogen ions, causing a drop in pH, but also reduces the concentration of carbonate ions (this happens because they react with hydrogen ions, forming bicarbonate). The problem for calcifiers is that carbonate ions are the basic building blocks they use to produce their shells. Many species need seawater to be supersaturated with carbonate ions to be able to form enough calcium carbonate for their skeletons and shells. As the concentration of carbonate ions drops, and seawater becomes undersaturated, calcifiers must devote more energy to pumping ions around their bodies and maintaining the process of shell-making. Molluscs have to concentrate carbonate ions in the gap between their mantles and their shells where new shell material is made. This can drain energy away from other vital functions, like reproduction and growth.

To make matters worse for shell-making molluscs, carbon dioxide also diffuses from water directly into their bodies, mostly through their gills. Left unchecked, a drop in the pH of body fluids can impact all sorts of important processes, in particular the functioning of enzymes. These proteins govern reactions around the body and they work best within a narrow pH range and will slow down or even stop if their surroundings become too acidic or too alkaline. As a consequence, organisms have evolved complex balancing mechanisms to maintain the right pH. Imagine a living body is a room, and acid-causing hydrogen ions are tennis balls that pour in through an open window; to prevent the tennis balls filling the room, and lowering the pH, you have to push them back out through the letterbox. Living bodies have various ways of keeping pH in balance, but they require yet more energy.

Lots of studies have tested how all sorts of calcifiers respond to falling pH and falling carbonate ion concentration. Coral reefs are a major focus for these studies because various components of these important tropical ecosystems form carbonate skeletons. This includes hard corals, the 'bricks' that form a reef's foundations, together with encrusting coralline algae that cement the reef together. It's possible that corals may adapt to gradual acidification and survive, but it's easy to be pessimistic about the future of reefs. The combined impacts of overfishing, coastal pollution, acidification and warming seas (which cause corals to lose the colourful, microscopic algae in their tissues, bleaching them white and in many cases killing them) lead many experts to think that coral reefs as we know them could be extinct by the end of the century.

Molluscs have also been the subject of extensive acidification research, in part because the valuable seafood industry could be left in ruins if edible species start disappearing. Clams, mussels, conchs, scallops, oysters and many more have all been plucked from their salty homes, moved into laboratory aquariums and exposed to seawater at various pH and carbon dioxide levels while scientists watch to see what happens. Initially, researchers mostly used mineral acids to simulate ocean acidification. They now tend to bubble carbon dioxide through water to more accurately mimic the real world. In most studies, as pH drops, the molluscs get in all kinds of trouble.

Flimsy and misshapen shells and lower calcification rates (the laying down of new shell material) are commonly seen in molluscs kept in seawater of lower pH and higher carbon dioxide than they're used to. Mussel byssus threads lose their stickiness, and many molluscs suffer from a suppressed immune system. Some researchers have observed molluscs swimming and crawling more slowly in acidified waters. Embryos and juveniles seem to be especially vulnerable. They take longer to mature and many don't survive.

It follows that with all the demands on their energy supplies, molluscs commonly respond to acidifying waters by boosting energy production or metabolic rate. They need energy to grow, to patch up shell damage, to try desperately to maintain their pH balance and ultimately to stay alive. For many species, all of these demands can become too taxing and they suffer, but this isn't always the case. Lab studies of ocean acidification regularly throw up unexpected and contradictory results.

Some molluscs seem quite unfazed by lowering pH and rising carbon dioxide, and some positively thrive. It seems to depend partly on where in the world the molluscs come from. The Blue Mussel is one species that confounds scientists by behaving differently in acidification studies around the world; in some places they are robust, elsewhere they do badly. It suggests that there is some degree of local adaptation to varying baseline conditions – pH is not the same everywhere in the oceans – and that some populations could be more likely to survive than others.

Slipper Limpets are another odd species. They have continued to grow happily when carbon dioxide levels around them were ramped up to 900 parts per million (or ppm; currently, the atmosphere is around 400ppm). When Common Cuttlefish are exposed to carbon dioxide at a massive 6,000ppm, some individuals remain unaffected and some actually do better than others kept in normal, mild conditions. After six weeks at extreme carbon dioxide levels, their internal cuttlebones, made of calcium carbonate, are bigger and heavier. Cephalopods, including these cuttlefish, are generally thought to have more sophisticated internal balancing mechanisms than other molluscs. They are also good at boosting their metabolism when they need to, which goes some way to explaining why cuttlefish get on so well in such extreme conditions. But plenty of puzzles still remain.

As it stands, the prognosis for shelled molluscs in acidifying oceans is mixed. Some species may be able to tough it out, while others will come to grief. But for sea butterflies in particular the prospects aren't looking too good.

Sea butterflies have been labelled the 'canaries in the coal mine' of acidifying oceans. These sensitive creatures could be the sentinels, warning of dangers ahead. In the first half of the twentieth century, miners would take down caged birds with them to detect toxic gases, mainly carbon monoxide; when the birds passed out and died, the miners knew it was time to put on breathing apparatus and make a quick escape. For sea butterflies in acidifying seas, the coal-mine analogy is rather ironic – or perhaps poetically dismal – seeing as it was coal that kick-started the problem of ocean acidification in the first place.

With their dainty, thin shells it comes as no great surprise that sea butterflies are among the more sensitive molluscs; they don't have much to lose shell-wise in the first place, so when exposed to acidifying waters they are especially vulnerable. Dire predictions suggest that swathes of the ocean could be out of bounds for sea butterflies in the years ahead.

Problems are likely to be most severe in polar seas, where acidification is expected to hit soonest and hardest, because cold water naturally holds more carbon dioxide. In parts of the Arctic and Antarctic, it's predicted surface seawater could become undersaturated with carbonate ions – and corrosive to unprotected shells and skeletons – within the next few decades. These frigid seas are also important parts of the sea butterflies' domain.

It's not easy to gauge how sensitive sea butterflies are to falling pH and rising carbon dioxide because they're flighty in more ways than one: they're especially tricky to keep alive in captivity. No one has yet worked out how to breed them, and they can't be shipped between labs around the world, so the only way to study them is to go to where they are in the wild.

The search for sea butterflies lured Silke Lischka deep inside the Arctic Circle. Among various research trips, she spent a winter in almost perpetual darkness illuminated from time to time by the Northern Lights. She was in Kongsfjord in the Svalbard archipelago, halfway between Norway and the North Pole, where a purpose-built research station makes it possible for scientists to live and work quite comfortably in this remote outpost. While she was there, sea butterflies were not difficult to find. Swarms of them would drift into the fjord and hover in the water right in front of the research station, and some days she could have sat on the dock and scooped them up in a bucket, but usually she puttered out in a boat and gathered her samples from deep fjord waters.

These were *Limacina helicina*, a close relative of the spiralling sea butterflies we found in Gran Canaria, and one of only a handful of sea butterfly species that live in the Arctic. After hatching over the summer, the juveniles have to survive a long, dark, hungry winter, hanging on until the sun returns in the spring, when they mature into adults, mate and produce the next generation.

With extreme care, Silke carried the young sea butterflies back to her laboratory on the shores of the fjord and kept them in a range of temperatures and carbon dioxide levels, including those levels expected by century's end. Then she measured their shells and examined them through a microscope for signs of damage.

At higher carbon dioxide levels, the transparent shells became more scuffed, perforated and scarred compared to those kept in more normal conditions; the high carbon dioxide sea butterflies were also slightly smaller, suggesting they weren't growing so well. The sea butterflies she hit with the combination of higher carbon dioxide levels and higher temperatures often didn't survive.

Repeating her experiments with empty shells, Silke showed that living sea butterflies can resist acidification to some extent; they don't get as badly damaged as empty, dead shells. But there's no doubt that having to reinforce their dissolving homes, laying down more carbonate on the inside, puts a strain on the juveniles' limited energy reserves. If this happened in the wild, the little sea butterflies would probably find it much harder to survive the winter.

Several other researchers have studied sea butterflies and uncovered similar gloomy forecasts of their demise. Clara Manno investigated sea butterflies in the far northern reaches of Norway. Her experiments showed not only that lower pH and higher carbon dioxide causes sea butterfly shells to lose weight, but she also revealed the confounding effect of freshwater. As sea ice and glaciers melt in a warmer world it's expected that the salinity of surface seawaters will drop. When both pH and salinity were reduced, sea butterflies flicked their wings more slowly as they swam around Clara's laboratory tanks, showing her that something was not right.

Like Silke, Steeve Comeau studied *Limacina helicina* in Svalbard, and found similar results, but he also ventured west to the Canadian Arctic, where he lived in a temporary research base perched out on the sea ice. Steeve collected his samples by lowering plankton nets through a hole in the ice. Back in the lab he found that the rate of calcification dropped by around 30 per cent in sea butterflies exposed to the carbon dioxide levels predicted by 2100.

In distinctly warmer waters, Steeve worked with larvae of a Mediterranean sea butterfly species. As he reduced pH, the larvae grew smaller, malformed shells. And he found that below pH 7.5, they didn't grow shells at all – but they didn't die. In the confines of the laboratory the naked sea butterflies seemed to get along just fine but there's no knowing if they would survive in the wild. Nobody has yet found any naked sea butterflies in the oceans, but one research team has uncovered the next worrying part of the story: wild sea butterflies whose shells already seem to be dissolving.

In parts of the oceans, winds blowing across the sea surface cause deep, cold waters to upwell into the shallows. These deeper waters are naturally rich in carbon dioxide and undersaturated with carbonate ions. Nina Bednaršek has led studies of sea butterflies in two upwelling regions. The first, in 2008, was in the Scotia Sea that stretches between Tierra del Fuego, at the tip of South America, and the island of South Georgia in the Subantarctic. The second, in 2011, was along the western seaboard of North America, between Seattle and San Diego. At both sites, Nina found sea butterflies with signs of damage and shell decay similar to those seen in animals that have been through acidification experiments in labs. Her findings have been interpreted as a worrying sign of things to come.

Will it matter if sea butterflies start to disappear from the oceans? Will declines or shifts in their range send ripples of change through the rest of the open ocean ecosystems?

One way that a loss of sea butterflies would potentially matter is because they play a part in drawing carbon away from surface seas down into the deep and away from the atmosphere. They do this via carbon locked up in organic matter, mostly their faeces.

Clara Manno was the first to identify sea butterfly droppings. They are compact pellets, oval in shape, greenish brown and quite easy to spot once you know how. She calculated that a single sea butterfly produces around 19 droppings per day, and they sink rapidly through the water column. Sifting through sediment samples gathered from the Ross Sea off Antarctica, she calculated that almost a fifth of all the organic carbon sinking into the depths – the so-called organic carbon pump – consisted of pteropod poo. Add their abandoned mucous webs, plus their dead bodies that get dragged down by their shells, and it means sea butterflies could drive half of the organic carbon pump in some polar waters. It's very difficult to predict exactly how things would change if sea butterflies were to begin abandoning acidifying waters in the Arctic and Antarctic. Other planktonic species could conceivably move in and take their place in the ecosystem, but there is always the chance that they would be less effective at removing carbon from the atmosphere and pulling it into the deep sea. If the organic carbon pump were to weaken, it would add yet another twist to the tangle of problems caused by climate change.

Without sea butterflies, there would also be a lot of hungry sea angels out there, as they eat little besides sea butterflies. Seabirds and fish also eat sea butterflies (although not exclusively); they in turn are eaten by bigger fish, as well as whales and seals, making sea butterflies a potentially crucial link in ocean food webs, including ones in which people are involved; there's a series of short hops from plankton to sea butterfly to salmon to dinner plate. If sea butterflies vanish or shift their ranges, it's possible the animals that eat them will also have to move, or find something else to eat, or go hungry. Exactly how important sea butterflies are as food for other animals, and whether ecosystems would be disrupted without them, is not clear. Much more research is needed.

It's true that sea butterflies can be extremely abundant and, when they are, other animals will often zero in and stuff themselves. Silke described to me a day during her time in Svalbard when a huge flock of sea butterflies drifted into the fjord; hundreds of kittiwakes and fulmars sat on the sea, merrily picking at the submerged feast. Other researchers have counted 10,000 sea butterflies in a single cubic metre of water, but such high densities only occur in patches that come and go.

A major challenge that lies ahead for ocean acidification research will be to move on from single-species studies. It's all very well knowing how individual animals react when exposed to acidifying seawater, but what happens when hundreds and thousands of organisms are all interacting, eating each other and competing for space and food? There's one thing everyone agrees on when it comes to understanding the impacts of ocean acidification: it's complicated. And most complicated of all will be predicting how entire ecosystems are going to respond. But researchers are finding ways.

How to probe an ecosystem

The departure lounge at Gran Canaria's Las Palmas airport overlooks the runway, and beyond it the Atlantic stretches out to the horizon. For two months in the late summer of 2014, if passengers glanced up from their Starbucks coffee and gazed through the huge glass walls, they might have caught a glimpse of science in progress.

Nine orange structures nod gently in the sea. Each is a ring of floating pipes sticking up into the air and supporting the top end of a giant, tubular plastic bag; two metres (six feet) wide and 15 metres (50 feet) long, it hangs down into the water. Umbrellas keep the rain off, and rows of spikes stop birds from landing and pooping on them. Down on the seabed, piles of iron railway wheels are used as anchors to hold the equipment in place. These structures act as giant test tubes, designed to test the effects of ocean acidification, not just on single species but on the profusion of life that makes up an open ocean ecosystem.

There's a bunch of clever things about these test tubes, which go by the name of KOSMOS, or the Kiel Off-Shore Mesocosms for future Ocean Simulation (mesocosm simply being a larger version of a microcosm, an encapsulated miniature world). For starters they are portable; they can be taken apart and shipped around the world, to repeat experiments in different sites, although this doesn't come without its challenges. In Sweden, the KOSMOS tubes were frozen in by sea ice and the previous spring, in Gran Canaria, some were torn to shreds by huge waves whipped up in a storm.

The KOSMOS tubes also benefit from being very big. To do something like this on land would be laborious and far more expensive; it would involve building huge tanks and pumping in seawater, causing who knows what confusion and damage to minute sea life in the process. Much better to take the test tubes to the ecosystem, rather than the other way round. The sides of the giant plastic bag are carefully lowered to enclose 55,000 litres of seawater and everything in it. Then the stage is set to manipulate conditions inside the tubes, in this case to pump in carbon dioxide at varying concentrations, to mimic the effects of ocean acidification.

Once that's done, the contents of the tubes are sampled every day or two. Water samples are extracted from the water column in each tube, traps at the bottom catch sinking particles and plankton nets are dragged through. Sampling all nine tubes can take hours, out in the dazzling, subtropical sunshine, but the really hard work has still to begin.

Back in the PLOCAN labs, the samples are divided up between researchers who eagerly whisk them off and plug them into an array of complex analytical devices, incubators and microscopes. By the time I pay them a visit, the 40-strong KOSMOS team has already had a few weeks to smooth out the kinks in their protocols but, even so, I'm amazed at how seamlessly the whole project is running.

Everyone knows what they're doing and the order in which things need to happen, as if they are part of their own well-functioning ecosystem. And what's more, they are all still smiling despite the long hours, roasting air temperatures, questionable coffee dispensed by the machine in the corridor and, for many of them, the repetitive, mind–numbing tasks – like counting sea butterflies.

Silke Lischka's job, along with her assistant Isabel, is to sort through all the debris caught in the sediment traps at the bottom of the mesocosm tubes and, as she and I had done, scour the plankton net samples. They do things a little more systematically, though; each of them has a counting chamber made from clear resin block with a long, narrow groove in it, the same width as the microscope's field of view, into which they pour the samples. They work their way along this elongated drop of water, counting sea butterflies as they go. Imagine an underground train driving slowly past while you stand on the platform counting all the people inside; you're much less likely to miss anyone, or count twice, compared to standing by a crowded swimming pool and doing a head count. A tally of sea butterflies is kept using an old-fashioned, mechanical counter with typewriter keys that clack when they're pressed and give a satisfying ding when they reach 100.

To get through all the samples takes hours, glued to a microscope, sometimes through long, sleepless nights. But I get a strong sense that everyone involved, especially Silke, knows why this is all worthwhile. They are contributing their part to a big, complex picture, probing the ecosystem from top to bottom, from the uptake and use of nutrients and the release of gases into the air, to viruses, phytoplankton and zooplankton. It's too early to say how the experiment is going, how the sea butterflies and all the other parts of the ecosystem are responding to different carbon dioxide levels, but by the end of the project, and following a great deal of data-crunching, the team will take a step back and trace a labyrinth of invisible connections.

Similar studies have been carried out already in Arctic waters and off the coast of Scandinavia, but this is the first time the KOSMOS mesocosms have been deployed in open seas. Beyond the edge of continental shelves, these clear, blue, nutrient-poor waters are representative of what two-thirds of the oceans look like. Understanding what happens here is a major part of predicting how life across the planet will respond to ocean acidification.

Head of the KOSMOS project is Ulf Riebesell from GEOMAR Helmholtz Centre for Ocean Research Kiel in Germany. I catch up with him after he has stayed up all night working on the latest stage of the experiment. The idea is to simulate the upwelling events that regularly take place when a steady current sweeping in from the north stirs up eddies in the island's wake, drawing deep water to the surface. The team used an enormous plastic bag to collect 80,000 litres of seawater, weighing 80 tonnes, from seven miles offshore and 650 metres down; the collecting bag took three hours to reach the surface, where it bobbed like a bloated whale. After they were set back a day by a broken water pump, the deep water was injected into the mesocosm tubes and all finally went according to plan. Now the team are on standby, waiting to see what effects unfold. They expect the nutrient-rich deep water will kick-start a phytoplankton bloom, and with it a feeding frenzy that will sweep through the rest of the ecosystem.

'It's like a big rain shower over a desert,' is the way Ulf describes the upwelling event to me. He is still wide awake and brimming with enthusiasm when we sit down to chat about the project. There is one thing in particular I want to ask him about, something that has been bothering me for a while: the passing of time.

Why time matters
Most ocean acidification studies take place over the course of hours and days and a few, like KOSMOS, keep going for months. But out in the real world, there is a hundred years to go before ocean pH is expected to reach extremely low levels. In that time, will marine life be able to adapt to the creeping changes in the world around it?

This is a major limitation of most ocean acidification studies; critics point out that they take place too fast, and don't continue for long enough, to truly mimic acidification in the oceans. A few longer-term studies hint that there is scope for adaptation, but perhaps only up to a point. Ulf's research group has bred phytoplankton called cocco-lithophores for 1,800 generations in high carbon dioxide conditions. These microscopic algae live inside clusters of calcium carbonate discs – collectively called the coccosphere – making them likely victims of acidification. However, over time, the laboratory population became more robust to falling carbonate saturation and lower pH.

The experiment acted as a form of artificial selection. The high carbon dioxide treatment slowed the growth rate of some coccolithophores, probably because they needed more energy to keep building their carbonate skeletons. Meanwhile some of them were more robust and were able to maintain their growth rates, perhaps even growing faster. These individuals were the ones that reproduced more rapidly, passing on more of their genes to the next generation. Slowly, in laboratory conditions, the acidified coccolithophores became adapted to their shifting water chemistry.

It remains unknown exactly how the coccolithophores' physiology changed; it's possible that as generations went by, they became more adept at ramping up metabolic rates and pumping ions around to maintain pH balance. Or, there may be some other as-yet unidentified mechanism that allows them to survive.

Could other calcifiers adapt to acidifying waters like coccolithophores? To repeat the experiments on anything that lives longer than these microbes would take an insanely long time. Coccolithophores have a generation time of a single day. It took five years to study them for 1,800 generations. For organisms like sea butterflies that have generations lasting a year, these experiments become quite unthinkable. Plus, it's well known that organisms with short generation times evolve quickly, compared to species that take longer to mature and reproduce (this is because the genomes of short-lived species are copied more frequently and errors quickly build up in their DNA, leading to more genetic variation that natural selection will act on). Coccolithophores are also highly abundant, with up to 10 million of them in a litre of seawater. It means that cocco-lithophores are inherently more adaptable to environmental changes than larger, rarer species like sea butterflies. And like Steeve Comeau's naked sea butterflies, the big unknown is whether carbon-resistant coccolithophores would survive out in the oceans, where there are masses of other species all competing for resources and space.

Even those organisms that can change their ways and adapt to a high carbon dioxide world may eventually still lose out. As the oceans continue to acidify, the cost of concentrating carbonate ions and building skeletons and shells will keep on steadily rising until calcifiers can simply no longer afford to make their homes.

'There are certain limits you can't pass,' Ulf tells me.

The century of acidifying seas that lies ahead will be unavoidably long and slow compared to the short-term studies aiming to forecast the future (after all, the only way to really know how the oceans are going to respond is to sit back and watch what happens in real time, but that's hardly the point of studies like this). However, the rate at which ocean acidification is now taking place is a mere beat of a sea butterfly's wings compared to the millennia that rolled by in previous climate change events, the ones that came and went before modern humans showed up. Sceptics point to these past events, to times when carbon dioxide levels were naturally high without humanity's input. *And look – look at all those things that were alive back then and are still here now.* There are still corals and plankton and all those molluscs with shells. They didn't melt away before, so why should we believe that will happen this time?

Things are different now. Given enough time, and a slow enough pace of carbon enrichment, the oceans themselves respond to ocean acidification and lessen its effects. Deep down on the sea floor there are vast deposits of calcium carbonate sediments, made from the fossilised remains of calcifying creatures – mostly coccolithophores and foraminifera – that lived and died over millions of years. In the past, carbon dioxide levels have risen in the atmosphere as is happening now, but from other sources besides human activities. The pH of shallow seas fell and, over the course of many centuries, those surface waters sank down, until they reached the deep carbonate sediments, causing them to dissolve and release carbonate ions. It meant that levels of carbon dioxide in the atmosphere were decoupled from the saturation of carbonate ions in the seas; while atmospheric carbon dioxide increased, it didn't drag down carbonate saturation with it. In essence, the oceans had their own colossal mechanism that buffered against acidification, which explains why many creatures with chalky skeletons were able to survive previous climate change events. In the past, calcifiers were protected from acidification by their ancestors – calcifiers of earlier eons – whose remains accumulated on the seabed. But now that link has been broken. The problem is, the oceans' inbuilt balancing mechanism takes 1,000 years or more to work, because that's how long it takes shallow waters to spread through the deep ocean. This time around, we don't have 1,000 years to wait.

Anthropogenic climate change is taking place much faster than anything the planet has experienced before, and the oceans can no longer keep pace with carbon emissions. The rate of uptake of carbon dioxide into the oceans far outstrips their ability to buffer against falling pH. Now, carbon dioxide levels and carbonate saturation are locked in relentless decline; side by side they drop together. The oceans today are slaves to the atmosphere.

A major talking point for climate change – and a target for sceptics – is the issue of how much experts agree on the facts. Increasingly, scientists worldwide are standing up and making it abundantly clear that they do agree, by and large, on the causes of climate change and the global troubles that could lie ahead. On a similar note, do experts agree that ocean acidification is happening, and that it's a problem for the seas? A 2012 survey of experts suggests that consensus is strong, at least when it comes to the bigger issues.

Jean-Pierre Gattuso from the Laboratoire d'Océanographie in Villefranche, France, led a survey asking 53 ocean acidification experts how much they agreed, or disagreed, with a list of statements. Almost all the experts agreed – without question – that ocean acidification is currently in progress, that it's measurable, and that it is mainly caused by anthropogenic carbon dioxide emissions ending up in the oceans (many pointed out that in coastal waters other pollutants, such as excess nutrients, can also affect pH).

As is the scientist's prerogative, many respondents picked apart the questions being asked, pointing out problems with the wording: 'What do you mean by "most"?', 'What does "adversely affect calcification" mean?'

In many cases, they emphasised the lack of certainty, the lack of long-term studies and the variable responses of different organisms. Without more data, it's difficult to be sure how food webs and fisheries will fare in a more acidic world. However, experts did agree that calcifiers, with their chalky skeletons and shells, are the marine species most likely to lose out.

Experts are also largely agreed that the ocean–atmosphere system has momentum. Even if carbon emissions were eradicated tomorrow, the oceans would continue to acidify for centuries to come. As one scientist put it, 'This is physical chemistry … I don't think there is any other possibility.' Does this mean that ocean acidification studies, like the KOSMOS mesocosms, are simply casting predictions about a global experiment that will run on regardless of their findings, and regardless of how humans behave in the next few decades? If ocean acidification really is inevitable and unstoppable, maybe it doesn't help to wrap our minds around the reality of how bad it will get. Perhaps we are better off not knowing.

I don't think so. There's no avoiding the uncomfortable truth that the only way to limit ocean acidification and the other problems of climate change – to stop the situation from becoming utterly disastrous – is to make drastic cuts to escalating carbon emissions, and to do it now. Decision-makers need to see these predictions, based on the best available science, of what a future world will look like so they can understand what it is that we're losing, and why action must be taken. The same goes for the rest of us. For most people, most of the time, ocean life is out of sight and out of mind, but there are plenty of good reasons why we should all sit up and take notice, and start caring about these vital, hidden worlds.

I felt a sense of great privilege peering at those sea butterflies and the other planktonic creatures as they whizzed around their glass-walled world, oblivious of me watching them. It was as if I had been let in on some of the oceans' greatest secrets, but who knows how much longer they will all be there? Of those spinning specks of life, some will be winners and others losers in the lottery of warmer, stormier and corrosive seas. And the really frightening thing is that the problems of the oceans don't stop at carbon. We are fishing deeper and further from shore than ever before, plundering wild species and treading paths of destruction through fragile ocean habitats. Dead zones are proliferating; garbage is piling up, transforming the open seas into toxic, plastic-flecked soup. All these troubles and many more combine, acting in concert to worsen each other. It's easy to feel overwhelmed, and utterly helpless in the face of relentless bad news.

But the problems are not all far away, nor are they out of our hands. It matters what each one of us decides to do, what we choose to eat, what we buy and what we throw away. We have the power to lighten our impact on the blue parts of our planet. Curbing as many individual problems as possible will give the oceans a chance to rest, to recover and restore themselves, and resist the impacts of climate change. If we act now, there's hope that in the years ahead there will still be a wealth of wonders in the oceans; there will be food for millions of people, from nutritious bowls of clams to the indulgent treat of flinty, raw oysters; sea snails will sneak up on sleeping fish and scientists will probe their spit for new inspirations; each night, nautiluses will rise from the inky depths, as they have done for hundreds of millions of years; tiny snails will fly around the open sea, spin webs to catch their food and be chased by other flying snails that don't have shells, and octopuses that do. And there will still be beautiful shells washing up on beaches, where people will find them and wonder where they came from, and how they were made.

Epilogue

In the summer of 2014, Philippe Bouchet led a team of mollusc-hunters to Nago, an island off the coast of Papua New Guinea, which lies in a coral-dotted lagoon stretching between the Bismarck Sea and the Pacific Ocean. This spot lies towards the eastern end of the Coral Triangle, the place where there are more marine species than anywhere else on the planet. Throughout years of field trips – in Vanuatu, Madagascar, the Philippines and elsewhere – Bouchet and colleagues from dozens of countries have been honing their collecting techniques. Divers venture out both day and night, so as not to miss the nocturnal species; they drop sampling devices down to varying depths where, like layers of a forest canopy, different assemblies of animals are found; they search between the tides; they even check among the spines of sea urchins and the tube feet of starfish for parasitic snails that suck the echinoderms' bodily fluids. The teams have also begun taking snippets of tissue to preserve the animals' DNA so species can be identified from their genetic fingerprint.

Using this suite of meticulous methods, the collectors are probing deeper than ever into the world of molluscs. In particular, they are uncovering an incalculable trove of micro-molluscs. These weeny animals are truly the secret gems of the sea. They come in a riot of exquisite colours, like a jar of jellybeans, with neon spots and stripes, yellow, purple, green and red. There are tiny clams with a sweep of pink tentacles sticking out, snails with glassy, transparent shells and kaleidoscopic mantles that show through from underneath, and bivalves that unfurl their colourful mantles over their shells and crawl about on their feet as if they were gastropods. Almost nothing is known about these animals. We don't know what they eat, what eats them, what strange and useful molecules they might contain or where exactly they belong on the sprawling tree of molluscan life. There are so few experts who specialise in these minute species that many of the samples Bouchet and the team collect could remain for years on a museum shelf, found and logged but not fully identified. Taxonomists call these neglected species 'orphans'. Even greater mysteries remain to be unravelled, hidden in oceanic nooks where molluscs reside but no one has yet worked out how to get hold of them.

Divers from Bouchet's team went gathering molluscs along a vertical wall of coral that plunges into the Bismarck Sea off Nago, 1,000 metres beneath the waves. The wall is pockmarked with caves, inside which the divers found a tantalising array of previously unknown shells, all of them between one and five millimetres in size, and all of them empty and dead. No matter how hard they tried and how carefully they looked, they couldn't find a single living mollusc responsible for making these enigmatic shells. The animals probably live deep within the cracks of this towering wall. The only reason we know they exist at all is because their shells drop down into the hands of the diving scientists. And we can only imagine what else lives in there, out of reach and out of sight.

There are undoubtedly many molluscs that will only be found by teams of experts with finely tuned searching skills and specialist equipment, but you don't need scuba gear or microscopes or deep-diving submersible vehicles to have your own encounters with a host of curious shelled creatures. Next time you visit a beach or swim in the sea or even take a stroll somewhere a long way from the ocean, look out for the shells that are all around you. Then you can read the stories written into shells, the clues left here and there that tell you about the shell-maker's life.

How big was your shell when it was a baby? Follow the spirals of a gastropod inwards towards its middle; the innermost whorls, the smoothest part often with an obvious line around the edge; this is the shell that the young snail wore when it first hatched. In some bivalves you can also spy a smooth inner part, right next to the hinge where the two halves of their shell fit together.

Which way does your shell coil? Hold it, tip pointing down, and see if you have a common right-coiler or perhaps it's a rare sinistral specimen, one that may have found life difficult and sex an awkward, mismatched challenge.

The shape of your shell, its ornaments, crenulations, striations and spines, will tell you about its life; perhaps it is a flattened clam that lay on the seabed, or a screw-shaped gastropod that dug its way down. Or is it covered in prongs to lodge itself in the sand, to try to stop it from being swept away?

Take a close look at patterns drawn across shells, those notes-to-self written so they didn't forget where they were in their shell-making efforts. Is there a point where the regular pattern goes awry? Did your shell get broken or attacked? Did it survive and keep on growing, eventually getting back in line and continuing with its elegant, decorated spiral?

Did your mollusc pick up any hitch-hikers, while it was alive or after it vacated its shell? You might spot barnacles or bryozoans (both formerly thought to be molluscs), or hydroids like tiny fir trees, or worms living inside white, spiralling tubes.

You might find gastropod shells with an elongated notch where a long siphon stuck out, probing and tasting the water in search of prey. These were the hunters, and you will also find their victims. Shells with neat holes drilled in them are testament to the evolution of so many molluscan ways of hunting and dining, and the fact that they're not shy of eating each other. Some shells might have a ring etched in them but not quite a hole, the sign of an interrupted assault.

Once you've read these stories, you can either leave the shells behind or take a few home as a reminder of a day at the beach or a walk in the woods. And maybe you'll be lucky enough to come face to face with shells that still have living occupants. Down at low tide you might spot dog whelks laying eggs like swollen grains of rice on the underside of boulders. In a rock pool you might catch a starfish attacking a limpet and getting its tube feet stamped on, or a pair of hermit crabs fighting over their shells. Perhaps, in shallow water, you'll spy a scallop swimming past like a living castanet, a cockleshell hopping across the seabed or a razor clam swiftly and efficiently digging its way out of sight. And maybe you'll find a sea snail, or a pond or land snail, and let it creep along your finger for a moment, watch it glide along its silvery trail with minute waves of its singular foot, before putting it carefully back where you found it.

A Note on Shell-collecting

If you buy a shell, especially a nice, shiny specimen, you should know that it wasn't picked up already empty and abandoned on a beach. Plenty of shells are left behind by molluscs that died of disease, predation, old age or some other fate, but those ones don't stay pristine for long. They get bashed about by waves and colonised by other living, encrusting, boring things, and soon they lose their gloss and are in no fit state for sale. Chances are that your gleaming shell was taken from a living animal; it was collected and killed and its shell removed and sold into the shell trade, so ultimately you could buy it.

Killing animals for human use is nothing new, of course, especially molluscs. The problem is that compared to the molluscs we eat, much less is known about ornamental species; large, beautiful shells taken from tropical coral reefs, small decorative gems and other shells used for their mother-of-pearl are all traded in huge quantities around the world. Exactly how many shells are traded each year is unclear because data are not always available or complete. It's thought around 5,000 mollusc species are targeted for their shells, but there's very little information about the impact of that harvest. However, anecdotal reports from shell collectors and sellers suggest that in many countries the trade is affecting wild populations. In Kenya, Tanzania, India and the Philippines many mollusc species have shells that are smaller than they used to be, a strong indication that all is not well and the larger specimens have been depleted. Local supplies have commonly run low, and traders have been forced to import shells from elsewhere; if you buy a shell on holiday in Hawaii or Florida, it most probably came from Mexico or Asia.

Despite all this, it is unlikely that a marine mollusc species will ever go extinct just because of collecting for the shell trade, except possibly for a few unlucky species with very small ranges. There have, however, been local extinctions. Giant clams and tritons have been wiped out from parts of the Indo-Pacific. In the Western Visayas of the Philippines Windowpane Oysters were virtually eradicated when mechanical dredges and rakes were used to scrape them up from the seabed; now the Windowpane Oyster industry in the Philippines, which crafts the shells into ornaments, relies on imports from Indonesia.

On the whole, the ornamental shell trade is less of a conservation concern than the sprawling global commerce in other marine species, such as the trade in shark fins to make into soup, or seahorses for traditional Asian medicines. Nevertheless, collecting shells can and does leave its mark on the natural world.

The inevitability that shells will become increasingly rare was the assumption behind a 1980s enterprise called Rare Shell Investment Services. 'There is little that can go wrong when investing in a disappearing, rare commodity' was the company's appalling claim. They encouraged investors to sink their cash into mollusc species that they reckoned would become very difficult to find but collectors would still want to own. Rare Shell Investment Services doesn't seem to be operating today, but there are signs that some shell prices have risen, while others fell when more specimens showed up, as happened with the Glory of the Sea.

However, if you care more about protecting wild species than plundering them for a profit, it's very difficult to know what to buy. The trade in ornamental shells tends to be poorly regulated and managed compared to the trade in edible species. Some countries have legislation protecting certain vulnerable species, although those laws aren't always well enforced. There are no major aquaculture efforts underway to farm ornamental molluscs besides giant clams (which are mainly for the aquarium trade and to re-stock coral reefs). International trade in all the giant clam species is strictly regulated by CITES (the Convention on International Trade in Endangered Species of Wild Fauna and Flora), so technically their harvest and sale shouldn't damage wild populations. If you do want to buy a giant clam, dead or alive, shell or living animal, you should make sure it's been traded with all the necessary permits.

Nautiluses and any other big shells should definitely be avoided (in general the bigger an animal, the longer it lives, the slower it grows and the more vulnerable it is to being over-hunted, which is certainly the case for nautiluses). On the whole it's impossible to know where a shell came from and how it was caught; there are no eco-labels for ornamental shells certifying that they come from sustainable sources. Unless you care a whole lot about having commercially bought shells in your life, it's perhaps best to resist temptation and leave them all alone.

As for finding your own seashells on beaches, here things are much more in your control. Always stick to local regulations and ask around; there might be a limit on the daily number and minimum size of shells that can be collected, you might need a permit to gather particular species and collecting is often prohibited in protected areas or nature reserves.

It's also up to you to make as little impact on the environment as possible while you're shell-collecting, although most of this is common sense: carefully turn rocks back over after you've peered underneath; don't trample across delicate habitats; don't take every single specimen you find, and remember that it makes ecological sense not to take any living molluscs.

And for your own sake, watch out for the cone snails.

Glossary
A Word in your Shell-like

Aculifera A proposed grouping of molluscs including all the animals without single shells; includes the chitons, solenogastres and caudofoveates.

Ammonite A group of extinct cephalopods, mostly with coiled shells, that lived during the Triassic, Jurassic and Cretaceous periods. They form part of the ammonoid lineage that first evolved in the Devonian. Their fossils were often known as snakestones.

Anthropogenic Caused or produced by humans, e.g. anthropogenic climate change.

Aragonite A form of calcium carbonate that is about 1.5 times more soluble than calcite. Most adult mollusc shells are made of aragonite. Some are calcite. Some are a bit of both.

Belemnite An extinct group of cephalopods with internal, bullet-shaped shells. Their fossils have often been called thunderstones.

Benthic Anything belonging to the seabed.

Bivalve The class of molluscs with shells in two parts (often more or less equal in size) including clams, mussels, cockles and scallops.

Byssus A term used since the fifteenth century for the strong, stretchy protein fibres with a sticky pad at one end that bivalves secrete from their feet to fix to rocks or to the seabed. Cleaned, carded and spun, they can form golden threads known as sea-silk.

Calcite A form of calcium carbonate that is more stable than aragonite.

Calcium carbonate A white solid made from calcium, carbon and oxygen ($CaCO_3$). It is the main building material of molluscan shells, and comes in two main forms: calcite and aragonite.

Caudofoveate Molluscs, but not as you know them. An obscure class of worm-like, shell-free animals that live in soft sediments. Also known as Chaetodermomorpha.

Cephalopod The class of molluscs including octopuses, squid, cuttlefishes, argonauts and chambered nautiluses. Most of them have lost or reduced their shells.

Chiton The class of molluscs that have eight shell plates lined up across their backs. They generally live clamped tightly to rocks and can roll up in a ball in defence. Pronounced 'kai-ton'.

Coleoid The mollusc lineage containing mostly soft-bodied cephalopods, including octopuses, squid, cuttlefishes and the extinct belemnites. They first emerged in the Devonian, around 400 million years ago.

Conchifera A grouping of the major groups of molluscs with shells: the gastropods, bivalves, cephalopods, scaphopods and monoplacophorans.

Conchology The scientific study or collection of mollusc shells.

Gastropod The class of molluscs also known as univalves, with a single and often spiralling shell, including snails and slugs.

Harmful algal bloom A dense aggregation of phytoplankton that produce harmful toxins and when consumed by filter-feeding bivalves can lead to various nasty (and occasionally lethal) shellfish poisoning symptoms. Previously known as red tides, but in fact they can be red, green, purple or brown.

Malacology The branch of zoology dealing with molluscs.

Mantle The layer of soft tissue that generally covers a mollusc's body and secretes the shell (if it has one).

Monoplacophoran A small and poorly known class of deep-sea molluscs that were thought to be extinct until specimens were found in the 1950s. They have limpet-like shells and radial symmetry, with multiple pairs of internal organs.

Morphospecies A 'probable' species that appears different from others based on the way it looks (morphological characters) but has not been fully and formally identified.

Nacre The shiny layer of a mollusc's shell (usually on the inside). Also known as mother-of-pearl.

Nautilid (Nautilida) A lineage of shelled cephalopods that first evolved in the Devonian period around 400 million years ago, leading up to the living chambered nautiluses.

Nudibranch A group of shell-less marine gastropods also known as sea slugs. Their name means 'naked gills'. Pronounced 'nudie-brank'.

Ocean acidification A reduction in the average pH of the oceans as a consequence of atmospheric carbon dioxide dissolving into seawater. The oceans have already become 30 per cent more acidic in the last 200 years.

Pelagic Anything belonging to the realm of the open sea.

Periostracum A layer of protein that covers the outside of a mollusc's external shell.

Phylum (plural phyla) Major category within the living world. Examples include molluscs, arthropods (crabs, shrimp, etc) and annelids (segmented worms). In turn, each phylum is traditionally divided up into classes, then orders, families, genera and species.

Plankton Microscopic, aquatic drifting creatures including both phytoplankton (plants and algae) and zooplankton (animals).

Pteropods An informal term for the gastropod orders Thecosomata (sea butterflies) and Gymnosomata (sea angels). These are pelagic snails that 'fly' underwater with little wings.

Radula The mouthparts of most molluscs. They come in a huge range of shapes and arrangements, allowing molluscs to specialise in different diets, from general herbivory to highly specialised hunting.

Sacoglossan A group of shell-less sea slugs that specialise in sucking sap from algae and plants.

Scaphopods The small class of molluscs also known as tusk shells. They look like miniature elephant's tusks, and generally live buried in seabed sediments.

Sclerites Bristles found on some molluscs, including chitons, solenogastres, caudofoveates and the now-extinct *Wiwaxia* (although not everyone agrees *Wiwaxia* was a mollusc).

Solenogastres An obscure class of molluscs. Like caudofoveates they are wormy and shell-free. They live either on the surface of mud or on corals.

Spat A common term for settled bivalve larvae, especially oysters and mussels.

Spondylus A genus of bivalve generally deep red, orange or purple in colour and covered in long spines that attract encrusting organisms (sponges, seaweeds, etc). Also known as thorny oysters.

Taxonomy The branch of science that occupies itself with identifying and naming living things, and sorting out how they are all related to each other.

Wiwaxia A creature from the Cambrian period around 520 million years ago, discovered in the Burgess Shale fossils. Some experts consider it to be an early mollusc. It had no shell but was covered in scales and bristles.

Select Bibliography

Chapter 1 Meet the Shell-makers

Bouchet, P., Lozouet, P., Maestrati, P. & Heros, V. 2002. Assessing the magnitude of species richness in tropical marine environments: exceptionally high numbers of molluscs at a New Caledonia site. *Biological Journal of the Linnean Society* 75: 421–436.

Johnson, S. B., Warén, A., Tunnicliffe, V., Van Dover, C., Wheat, C. G., Schultz, T. F. & Vfrijenhoek, R. C. 2014. Molecular taxonomy and naming of five cryptic species of *Alviniconcha* snails (Gastropoda: Abyssochrysoidea) from hydrothermal vents. *Systematics and Biodiversity* 1–18.

Kocot, K. M. 2013. Recent advances and unanswered questions in deep molluscan phylogenetics. *American Malacological Bulletin* 31: 195–208.

Ponder, W. F. & Lindberg, D. R. R. 2008. *Phylogeny and Evolution of the Mollusca.* University of California Press, Berkeley.

Smith, M. R. 2014. Ontogeny, morphology and taxonomy of soft-bodied Cambrian 'Mollusc' *Wiwaxia. Palaeontology* 57: 215–229.

Chapter 2 How to Build a Shell

Boettiger, A., Ermentrout, B. & Oster, G. 2009. The neural origins of shell structure and pattern in aquatic mollusks. *PNAS* 106: 6837–6842.

Clements, R., Liew, T.-S., Vermeulen, J. J. & Schilthuizen, M. 2008. Further twists in gastropod evolution. *Biology Letters* 4: 179–182.

Gong, Z., Matzke, N. J., Ermentrout, B., Song, D., Vendetti, J. E., Slatkin, M. & Oster, G. 2012. Evolution of patterns on *Conus* shells. *PNAS Early Edition* DOI: 10.1073/pnas.1119859109

Hoso, M., Kameda, Y., Wu, S.-P., Asami, T., Kato, M. & Hori, M. 2010. A speciation gene for left–right reversal in snails results in anti-predator adaptation. *Nature Communications* DOI: 10.1038/ncomms1133

Meinhardt, H. 2009. *The Algorithmic Beauty of Seashells.* Springer, Dordrecht, Heidelberg, London & New York.

Raup, D. R. 1962. Computer as aid in describing form in gastropod shells. *Science* 138: 150–152.

Thompson, D'Arcy Wentworth. 1917. *On Growth and Form.* Cambridge University Press, Cambridge. Reprinted 1992.

Vermeij, G. J. 1995. *A Natural History of Shells.* Princeton University Press, Princeton.

Chapter 3 Sex, Death and Gems

Bouzzouggar, A., Barton, N., Vanhaeren, M., d'Errico, F., Collcutt, S., Higham, T., Hodge, E., Parfitt, S., Rhodes, E., Schwenninger, J.-L., Stringer, C.,

Turner, E., Ward, S., Moutmir, A. & Stambouli, A. 2007. 82,000-year-old
    shell beads from North Africa and implications for the origins of modern
    human behavior. *PNAS* 104: 9964–9969.
Claassen, C. 1998. *Shells.* Cambridge University Press, Cambridge.
Gaydarska, B., Chapman, J.C., Angelova, I., Gurova, M. & Yanev, S. 2004.
    Breaking, making and trading: the Omurtag Eneolothis *Spondylus* hoard.
    Archaeologia Bulgarica 8: 11–33.
Hogendorn, J. & Johnson, M. 1986. *The Shell Money of the Slave Trade.*
    Cambridge University Press, Cambridge.

Chapter 4 Shell Food

Diaz, R. J. & Rosenberg, R. 2008. Spreading dead zones and consequences for
    marine ecosystems. *Science* 321: 926–929.
Glibert, P. M., Anderson, D. M., Gentien, P., Granéli, E. & Sellner, K. G. 2005.
    The global, complex phenomenon of Harmful Algal Blooms. *Oceanography*
    18: 136–147.
Potasman, I. & Odeh, M. 2002. Infectious outbreaks associated with bivalve
    shellfish consumption: a worldwide perspective. *Clinical Infectious Diseases* 35:
    921–928.
Richter, C., Rao-Quiaoit, H., Jantzen, C., Al-Zibdah, M. & Kochzius, M. 2008.
    Collapse of a new living species of giant clam in the Red Sea. *Current
    Biology* 18: 1349–1354.
For online advice on making better seafood choices:
Monterey Bay Aquarium Seafood Watch, www.seafoodwatch.org
Marine Conservation Society Fishonline, www.fishonline.org
Australia's Sustainable Seafood Guide, www.sustainableseafood.org.au

Chapter 5 A Mollusc Called Home

Beck, M. W., Brumbaugh, R. D., Airoldi, L., Carranza, A., Coen, L. D.,
    Crawford, C., Defeo, O., Edgar, G. J., Hancock, B., Kay, M. C., Lenihan, H.
    S., Luckenbach, M. W., Toropova, C. L., Zhang, G. & Guo, X. 2011. Oyster
    reefs at risk and recommendations for conservation, restoration, and
    management. *Bioscience* 61: 107–116.
zu Ermgassen, P. S. E., Spalding, M. D., Grizzle, R. E. & Brumbaugh, R. D. 2013.
    Quantifying the loss of a marine ecosystem service: filtration by the eastern
    oyster in US estuaries. *Estuaries and Coasts* 36: 36–43.
Haires, D. 2013. The flame shells of Kyle Akin. *Mollusc World* 32: 15–17.
Kirby, M. X. 2004. Fishing down the coast: historical expansion and collapse of
    oyster fisheries along continental margins. *PNAS* 101: 13096–13099.
Laidre, M. E., Patten, E. & Pruitt, L. 2012. Costs of a more spacious home
    after remodelling by hermit crabs. *Journal of Royal Society Interface*
    DOI: 10.1098.
Lewis, S. M. & Rotjan, R. 2009. Vacancy chains provide aggregate benefits to
    Coenobita clypeatus hermit crabs. *Ethology* 115: 356–365.

Chapter 6 Spinning Shell Stories

Hendricks, I. E., Tenan, S., Tavecchia, G., Marbà, N., Jordà, G., Deudero, S., Álvarez, E. & Duarte, C. M. 2013. Boat anchoring impacts coastal populations of the pen shell, the largest bivalve in the Mediterranean. *Biological Conservation* 160: 105–113.

Maeder, F. 2008. Sea-silk in Aquincum: first production proof in antiquity. *Purpureae Vestes. II Symposium Internacional sobre Textiles y Tintes del Mediterráneo en el mundo antiguo* (eds C. Alfaro & L. Karali), pp. 109–118.

McKinley, D. 1998. Pinna and her silken beard: a foray into historical misappropriations. *Ars Textrina* 29: 9–223.

Project Sea-silk website: www.muschelseide.ch/en

Chapter 7 Flight of the Argonauts

Broderip, W. J. 1828. Observations on the animals hitherto found in the shells of the genus *Argonauta*. *The Zoological Journal* 4: 57–66.

Finn, J. K. & Norman, M. D. 2010. The argonaut shell: gas-mediated buoyancy control in a pelagic octopus. *Proceedings of the Royal Society B: Biological Sciences* 277: 2967–2971.

Hewitt, R. A. & Westermann, G. E. G. 2003. Recurrences of hypotheses about ammonites and argonauta. *Journal of Paleontology* 77: 792–795.

Kruta, I., Landman, N., Rouget, I., Cecca, F. & Tafforeau, P. 2011. The role of ammonites in the Mesozoic marine food web revealed by jaw preservation. *Science* 331: 70–72.

Landman, N. H., Goolaerts, S., Jagt, J. W. M., Jagt-Yazykova, E. A., Machalski, M. & Yacobucci, M. M. 2014. Ammonite extinction and nautilid survival at the end of the Cretaceous. *Geology* DOI: 10.1130/G35776.1

Chapter 8 Hunting for Treasures

Barord, G. J., Dooley, F., Dunstan, A., Ilano, A., Keister, K. N., Neumeister, H., Preuss, T., Schoepfer, S. & Ward, P. D. 2014. Comparative population assessments of Nautilus sp. in the Philippines, Australia, Fiji, and American Samoa using baited remote underwater video systems. *Plos ONE* 9: DOI: 10.1371/journal.pone.0100799

Dance, S. P. 1986. *History of Shell Collecting*. E. J. Brill, Leiden.

De Angelis, P. 2012. Assessing the impact of international trade on chambered nautilus. *Geobios* 45: 5–11.

Reeve, L. A. & Sowerby, G. B. 1843–1878. *Conchologia Iconica, or Illustrations of Shells of Molluscous Animals*. Lovell Reeve, London.

Chapter 9 Bright Ideas

Finnemour, A., Cunha, P., Shean, T., Vignolini, S., Guldin, S., Oyen, M. & Steiner, U. 2012. Biomimetic layer-by-layer assembly of artificial nacre. *Nature Communications* 3: DOI: 10.1038/ncomms 1970

Kohn, A. J. 1956. Piscivorous gastropods of the genus *Conus*. *Zoology* 42: 168–171.

Li, L. & Ortiz, C. 2014. Pervasive nanoscale deformation twinning as a catalyst for efficient energy dissipation in a bioceramic armour. *Nature Materials* 13: 501–507.

Mirkhalaf, M., Dastjerdi, A. K. & Barthelat, F. 2014. Overcoming the brittleness of glass through bio-inspiration and micro-architecture. *Nature Communications* DOI: 10.1038/ncomms4166

Olivera, B. M. & Cruz, L. J. 2001. Conotoxins, in retrospect. *Toxicon* 39: 7–14.

Peters, H., O'Leary, B. C., Hawkins, J. P., Carpenter, K. E. & Roberts, C. M. 2013. *Conus*: first comprehensive conservation Red List assessment of a marine gastropod mollusc genus. *Plos ONE* 8: DOI: 10.1371/journal.pone.0083353

Seronay, R. A., Fedosov, A. E., Astilla, M. A., Watkins, M., Saguil, N., Heralde III, F. M., Tagaro, S., Poppe, G. T., Aliño, P. M., Oliverio, M., Kantor, Y. I., Concepción, G. P. & Olivera, B. M. 2010. Biodiverse lumun-lumun marine communities, an untapped biological and toxinological resource. *Toxicon* 56: 1257–1266.

Winter, A. G., Deits, R. L. H., Slocum, A. H. & Hosoi, A. E. 2014. Razor clam to RoboClam: burrowing drag reduction mechanisms and their robotic adaptation. *Bioinspiration & Biomimetics* 9.

Yao, H., Dao, M., Imholt, T., Huang, J., Wheeler, K., Bonilla, A., Suresh, S. & Ortiz, C. 2010. Protection mechanisms of the iron-plated armor of a deep-sea hydrothermal vent gastropod. *PNAS* 107: 987–992.

Chapter 10 The Sea Butterfly Effect

Bednarsek, N., Feely, R. A., Reum, J. C. P., Peterson, B., Menkel, J., Alin, S. R. & Hales, B. 2014. *Limacina helicina* shell dissolution as an indicator of declining habitat suitability owing to ocean acidification in the California Current Ecosystem. *Proceedings of the Royal Society B: Biological Sciences* 281.

Caldeira, K. & Wickett, M. E. 2003. Anthropogenic carbon and ocean pH. *Nature* 425: 365.

Comeau, S., Gorsky, G., Alliouane, S. & Gattuso, J.-P. 2010. Larvae of the pteropod *Cavolinia inflexa* exposed to aragonite undersaturation are viable but shell-less. *Marine Biology* 157: 2341–2345.

Gattuso, J.-P. & Hansson, L. 2011. *Ocean Acidification.* Oxford University Press, Oxford.

Gattuso, J.-P., Mach, K. M. & Morgan, G. 2013. Ocean acidification and its impacts: an expert survey. *Climatic Change* 117: 725–738.

Gazeau, F., Parker, L. M., Comeau, S., Gattuso, J.-P., O'Connor, W. A., Martin, S., Pörtner, H. & Ross, P. M. 2013. Impacts of ocean acidification on marine shelled molluscs. *Marine Biology* 160: 2207–2245.

Lalli, C. M. & Gilmer, R. W. 1989. *Pelagic Snails: The Biology of Holoplanktonic Gastropod Mollusks.* Stanford University Press, Stanford.

Lischka, S., Büdenbender, J., Boxhammer, T. & Riebesell, U. 2011. Impact of ocean acidification and elevated temperatures on early juveniles of the polar shelled pteropod *Limacina helicina*: mortality, shell degradation, and shell growth. *Biogeosciences* 8: 919–932.

Acknowledgements

Turning my attention from a single, obscure genus with 40 or so species to an entire, globe-spanning phylum containing hundreds of thousands of motley creatures was, perhaps, a bold move. Writing about seashells and molluscs has been an altogether different experience compared to exploring the world of seahorses and luckily a lot of wonderful people have been there to help me navigate these broad, rambling reaches of the animal kingdom.

I am deeply grateful to all the researchers who have shared with me their molluscan enthusiasms and ideas, answered my questions and helped me make fewer mistakes than I would have on my own (any slip-ups that are still in the book are entirely down to me). Thank you to Philippe Bouchet, Martin Smith, Reuben Clements, Thor-Seng Liew, Bard Ermentrout, George Oster, Masaki Hoso, Bisserka Gaydarska, Dan Harries, Philine zu Ermgassen, Piero Addis, Vicky Peck, Nina Bednaršek, Gareth Lawson, Julian Finn, Ken McNamara and Baldomero Olivera.

It was a great honour to be granted a Roger Deakin award for this book from the Authors' Foundation at the Society of Authors. This gave me a link to one of my favourite and much-missed nature writers and allowed me to carry out a series of research trips. My hunt for sea-silk in Sardinia would not have been possible, or nearly as much fun, if it weren't for Alessandro Spiga and Silvia Messori, who so warmly welcomed me into their home, took me snorkelling to see Noble Pen Shells and introduced me to Chiara Vigo. Thank you also to Annelise Hagan and Eleonora Manca for putting me in touch with the people of Sant'Antioco, to Rebecca Lewis for coming along on our adventure and translating for me, and to Chiara for showing me her work. My sea-silk story would have been impossible without the kindness and knowledge of Felicitas Maeder, especially for introducing me to the people at Archeotur in Sant'Antioco. My thanks in particular go to Ignazio Marrocu, Giustino Argiolas and Patrizia Zara, and of course to Giuseppina and Assuntina Pes for inviting me into their home and demonstrating their sea-silk skills.

I am hugely grateful to Ulf Riebesell for inviting me to join him in Gran Canaria, and to the rest of the BIOACID team who kindly took me out to Gando Bay to see the KOSMOS mesocosms and let me snoop around their labs. A very special thanks goes to Silke Lischka for so graciously helping me find sea butterflies and sharing her immense enthusiasm for these tiny creatures when she really should have been sleeping and recovering from the gruelling research schedule.

In the UK, a big thank you to fellow Triton fan Andy Woolmer for showing me around the Mumbles, and for all his insights into oysters, whelks, cockles, mussels and the rest (and for persuading me to try winkles for the first time). Thank you to Jon Ablett for showing me behind the scenes at London's Natural History Museum, and to Peter Dance for our discussions, beginning several years ago, about Hugh Cuming, for sharing his Cuming archive with me, and for treating me to the best Thai clams I've ever tasted. A warm thank you to Fatou Janha and all the women of the TRY Oyster Women's Association in The Gambia. If you visit The Gambia, make sure you try the oysters.

This book wouldn't have happened without Jim Martin at Bloomsbury, who has been the ideal combination of editor and molluscan co-conspirator. Many thanks to him for indulging and sharing my shelly whims, and for being so utterly selfless in the face of many edible molluscs. Our journey to the book's cover and illustrations began when I spotted a beautiful drawing of an argonaut on Aaron John Gregory's website. When I discovered that Aaron is not only a talented artist but also as much of a marine geek as me, I instantly knew that he was our man. A huge thank you to Aaron for his immense patience and hard work, and for so brilliantly bringing the molluscs to life.

Lastly I want to thank all my dear friends and family who have cheered me on through my seashell adventures, who have read my words, sat through all the shell stories, and in many ways kept me going. My love and gratitude go to you all, and in particular to Anna Petherick, Riamsara Kuyakanon Knapp, Eric Drury, Matthew Wilkinson (whose book on animal locomotion was being written at the same time as this one), Ria and Jake Snaddon (plus baby Snaddon who will arrive in the world shortly before this book does, and who I look forward to showing seashells in the years ahead), Peter Wothers, Umut Dursun, Conor Jamieson, Liam Drew, Joshua Drew, Drew Bednarski and Meghan Strong, Kate Lash (my official geochemistry consultant), and finally my parents, Di and Tom Hendry, my mum especially for coming up with the book's inspired subtitle, and my dad for reading so much of the manuscript when he should have been working on his Ph.D. And Ivan, my constant companion in life and words, who calls me up on my smutty jokes, finds ways to help me tell my stories, and always makes things better.

Index